The Future of Earth's Inhabitants: Based on the Past and Present

Glenn Bell, Ph.D.

First Printing: 2022

ISBN 978-1-7328379-4-2

Front Cover Photo

Picture of the Earth

europe-globe-ter-terrestrial-globe.jpg (910×512) (pxfuel.com)

Note: Use of this photo as well as all other figures in the book does not in any way suggest an endorsement of this book.

The Future of Earth's Inhabitants:
Based on the Past and Present

Table of Contents

Introduction

Change is seen over the passing of time. Typically, we observe changes over periods of time that we can observe in our lives such as over seconds, minutes, hours, days, months, or years. These are timescales that we can observe using the memory of our past. From information on astronomical and geological history we recognize that time has existed well beyond our personal observation, and we recognize that there was a vast time before humans and before life on Earth. This historical information is derived from a combination of writings, artifacts, fossils, geological evidence, measurements, and scientific principles.

We also recognize that there is vast time in the future, much greater than all the time in the past. We largely make plans in time spans that are again associated with our life spans. Occasionally, people will make plans for developments in the more distant future. For example, some building projects have continued over multiple life spans (i.e., the Great Wall of China (2000 years), Chichen Itza (400 years), Angor Wat (400 years)). Some people may make plans for dealing with issues in the distant future, like storage of fission reactor waste, which will last for millions of years. These plans are made, based on expectations with an imperfect knowledge of how the future will occur.

If we look over time, we see that there has been a sequence of important events including the origin of the Universe, generation of the planets including the Earth, the origin of life, and the evolution of life forms including the origin of humankind. Most of these events have occurred millions or billions of years in the past. It is the purpose of this book to examine the likely futures.

To look at the future of the Earth's inhabitants, we will look at the past and then we can potentially make predictions for the near and distant future. In this book, I use the terms Earth's inhabitants or

Earthlings to refer to life forms that come from Earth, which are generally believed to have originated on Earth. While it is possible or probable[i] that life exists elsewhere in the Universe, we will discuss the future of life that is currently present on Earth. If there is life elsewhere in the Universe, it is unclear whether such lifeforms would be similar to our lifeforms, due to potential differences in environmental forces on other planets, albeit they would have to operate according to the same universal laws of physics.

Of special interest to us is humanity's future. While it is evident that some present life forms will become extinct and others will evolve, we will look at anticipated changes. We will also try to predict whether Earthlings will spread to other planets, or become extinct due to war, disease, or for other reasons.

While no one can foretell the future for certain, it may be possible to predict some future events based on knowledge of scientific principles, history, and based on trends. One challenge to prediction is biases that are based on a limited perspective; so, we will try to look at things from many perspectives and have an open mind.

Predicting the Future[ii]

Scientific principles are the soundest information to use in predicting the future. When we understand measured patterns of behavior, such as the laws of motion[iii] and gravitational forces, we can predict that if we drop an object it will move toward the center of the Earth. We can predict motions reliably and can understand cases in which other forces than that of gravity could alter the prediction. While scientific principles are the best information enabling prediction, our understanding of science continues to change and improve, which is fundamental to the scientific process. Improved understanding of the nature and behavior of the Universe may lead to major impacts in both scientific and technological developments. We will use our current knowledge of the principles of physics (including nuclear physics and planetary motion), geology, biology, and evolution in anticipating the future.

Geological, biological, and human history provides many general principles related to recurring events that are cause and effect related. History, both written and unwritten, provides examples of the catastrophic events such as asteroid collisions, volcanic eruptions, earthquakes, and environmental changes that are difficult to predict but can be expected to reoccur. Other historical events such as extinctions and the generation of new species are recurring themes over history. Examining the rise and fall of kingdoms, empires, and civilizations can provide insight potential future changes. History provides us information on the behavior of bacteria, plants, and animals including humans and the consequences of these behaviors. For example, if there is unequal distribution of important resources such as water, food, or energy, we may be able to predict the actions of those life forms that are in need.

Trends are useful, but can change, so they should be considered carefully. We can identify current trends like those of an increasing human population, increasing use of energy, increasing the rate of extinction of species, increasing pollution, and increasing water shortages. We can speculate on whether or how these trends will change. We can also speculate on the potential impacts of these changes.

Technological advances will likely have a significant effect on the future. Some technological advances can be anticipated, while other advances that will occur are likely not conceived currently. We will speculate on some of the types of technological advances anticipated and their impact on the future.

Since the life span of any individual is short in relationship to the age of the Universe, geological changes, or evolutionary change, it is often difficult for us to look beyond ourselves and into the distant future. Often, we are more interested in concerns associated with the next, day, week, month, or year. This is practical but, in this book, we will look over a much broader time span. We will attempt to predict events in the near future, distant future, and near the end of time.

Chapter 1 - A Look at the Past

To enable us to predict the future, we should first try to understand the past. A look at the past provides insight into changes that may be expected to persist or repeat due to principles of science. It also provides a perspective on how we got to the current astronomical, geological, environmental, biological state and helps identify short-term and long-term trends.

Origin of the Universe

The first and most important event is the origin of the Universe. Scientists think that the origin of the Universe is best explained by evidence supporting the Big Bang theory. The Big Bang theory states that the Universe is the product of a singularity or a point of high density and high temperature in which particles and energy fluctuated in quantum fields (fields incorporating quantum mechanics and the principles of the theory of relativity). The explosion or expansion of this singularity is estimated to have occurred 13.8 billion years ago based partly on the current velocities of galaxies. This event generated time, space, energy, and matter. The figure of 13.8 billion years is based on identification of the oldest stars and determining their ages based on knowledge of how stars are born, evolve, and die. The Big Bang theory is supported by data regarding an expanding universe, primordial abundance of the light elements, and the existence and character of a pervasive, ordered microwave background radiation. An initial extremely brief period of superluminal (appearing to travel faster than the speed of light) expansion generated the largest scale structures of the galaxies that are currently observed.

Since then, the Universe has been tending toward disorder. This increase in disorder is explained by matter tending toward the most likely state, known as the second law of thermodynamics.

The second law of thermodynamics states that there is a natural tendency of an isolated system to degenerate into a more disordered state, which is a more probable arrangement. Based on observations of galaxies, the Universe is expanding, and all energy and matter is a product of the singularity. The observations show a red shift in the wavelengths of light. In this case, it is a red shift indicating that most objects are receding away from us. That is, light received from most stars has a frequency shift toward red (lower frequency), much like the tone of a siren coming from an emergency vehicle moving away from us has a low frequency shift. The galaxies are clustered with each other due to quantum fluctuations (temporary random change in the amount of energy in a point in space) in matter and energy in much less than a nanosecond after the initial Big Bang.

Within the Universe's first second, it cooled enough for the matter to coalesce into protons and neutrons, which are the particles that make up an atom's nuclei. As the Universe expanded in the first minutes, the protons and neutrons assembled into hydrogen and helium nuclei. By mass, hydrogen was around 75 percent of the early Universe's matter, and helium was around 25 percent[iv]. Despite having atomic nuclei, the Universe was too hot for electrons to form stable atoms. The Universe's matter remained an electrically charged fog. After about 380,000 years, the Universe cooled down enough for neutral atoms to form. There were no stars in the Universe until about 180 million years after the Big Bang. Matter coalesced through gravitational forces and by dissipating energy through radiation to form stars and planets, and galaxies.

The movement of particles in matter generates thermal radiation or heat. The radiation is spread over the electromagnetic spectrum. The peak and distribution of the emissions are a function of the temperature. As the Universe expands and cools, the emitted thermal radiation spectrum changes in peak and distribution. The distribution of thermal radiation is uniquely determined by the temperature. Background thermal cosmic microwave radiation was detected by Penzias and Wilson in 1960 at Bell Labs. More extensive information was obtained from the Cosmic Background Explorer (COBE) in 1989, which indicated that the cosmic radiation was consistent with a temperature of 2.736 above absolute zero, which is consistent with predictions based on the Big Bang theory[v].

Based on this information, we can expect the continued expansion of the Universe for some time, and we will revisit this later in the book (see sections on Collision of Andromeda Galaxy and Milky Way Galaxy and End of the Universe). We can also expect a continual increase toward disorder or increased entropy. It is also worth noting that the Universe has undergone drastic changes to get to our present status. Thus, it is possible that more drastic changes could occur in the future. We will look at this as we examine possible ends of the Universe.

Our Solar System and the Sun

About 4.6 billion years ago, our Solar System formed from a cloud of gas and dust, which slowly contracted under the mutual gravity of its particles. This was aided and hindered by pressures in the magnetic fields and light emitted from the matter itself, as well as by surrounding stars and their supernovae. The age of the Sun can be estimated based on the ages obtained from radioactive dating of the oldest meteorites, since the Solar System formed as a unit, within a span of tens of millions of years according to physical models and the tight spreads of determined dates. Based on radioactive dating[1], the oldest Earth rocks are calculated to be around 4.6 billion years old.

The nebula or cloud of gas and dust may have been the product of a supernova of a massive star. Massive stars have relatively short lives in comparison to smaller stars and are cable of generating

---

[1] Radioactive dating is a technique used to date matter that compares the abundance of naturally occurring radioactive isotopes within a material to the abundance of decay products. Radioactive isotopes have well known decay rates (half-lives) that can be used to mathematically determine the age of a sample. The half-life of an isotope is the time taken by its nucleus to decay to half of its original number.

heavier elements during the supernova explosion. If the star is sufficiently massive, the collapsing core may become hot enough to support nuclear reactions that consume helium and produce a variety of heavier elements up to iron. The cloud was made largely of hydrogen with some helium and small amounts of naturally occurring chemical elements. The matter moved in space due to gravity, leading to a rotation or tumbling. The initial rotation or tumbling motion was accelerated as the nebula contracted. The cloud became a disc. Within the disc, the largest concentration of matter was in the center. This concentrated matter became our Sun.

Matter collected in smaller clumps further out in the disc. These became the planetesimals, which are bodies that come together under gravitation to form a planet. The proto-sun and proto-planets grew by accretion of the matter that was moving in toward the center of mass as it lost energy by emitting light. The Solar nebula (a large cloud of gas and dust) became hotter as the contraction due to gravitational forces increased the pressure, especially in the inner portion of the nebula.

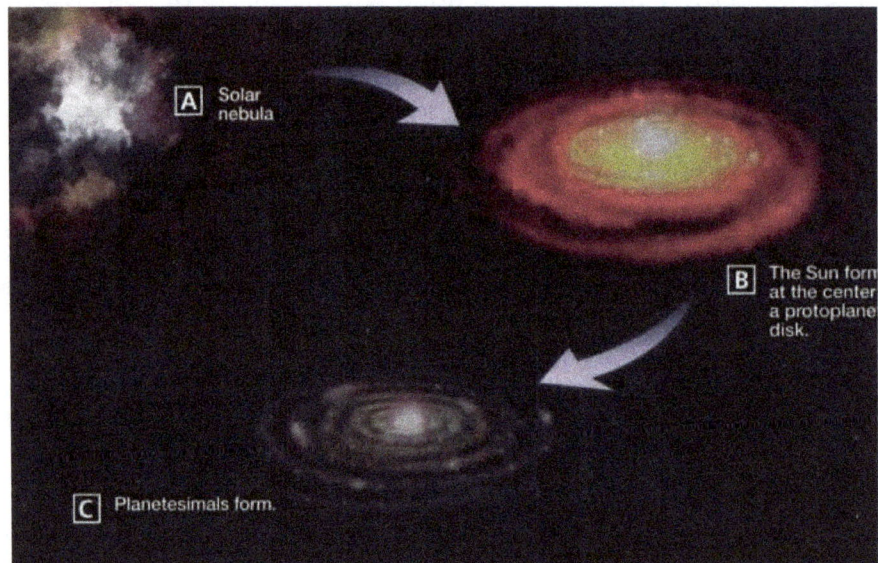

Origin of Our Solar System[vi]

The Sun is a large collection of hydrogen of sufficient mass to generate a nuclear reaction. Hydrogen gas is fused to make

14

helium through nuclear fusion. Nuclear fusion[vii] is a reaction in which two or more atomic nuclei are combined to form one or more different atomic nuclei and subatomic particles (neutrons or protons). The difference in mass between the reactants and products is converted into energy. The amount of energy generated is calculated using the following equation:

$$E = mc^2$$

where energy [E] is equal to mass [m] times the speed of light [c] squared.

This difference in mass arises due to the difference in atomic "binding energy" between the atomic nuclei before and after the reaction. Fusion is the process that enables stars to release energy into space. The atomic fuel of the Sun is continuing to be used to generate light, heat, and other electromagnetic energy.

The Sun will continue to release energy from the fusion reactions, losing mass in the process. The hydrogen is used as fuel, which will over billions of years become exhausted. We will revisit this in discussion of the evolution of the Sun later (see section on Consequences of the Evolution of the Sun).

The evolution of stars can be described by luminosities and temperatures in a graph known as the Hertzsprung-Russell diagram. If one can determine the stellar mass, then the temperature, luminosity, and radius can be estimated from a pattern along the main part of the diagram. About 90% of all stars fall on the diagonal, called the main sequence, between hot, luminous stars in the upper-left-hand corner and cool, dim stars in the lower-right-hand corner. During the time in which stars are found on the main sequence line, they are fusing hydrogen in their cores. The next concentration of stars is on the horizontal branch (helium fusion in the core and hydrogen burning in a shell surrounding the core). Importantly, massive stars produce the elements carbon, nitrogen, and oxygen, which are used by lifeforms on Earth. Our Sun currently lies near the center of the main sequence. The evolution of our Sun is expected to have begun as a protostar and now to the main sequence. It is expected to become a red giant and eventually a white dwarf.

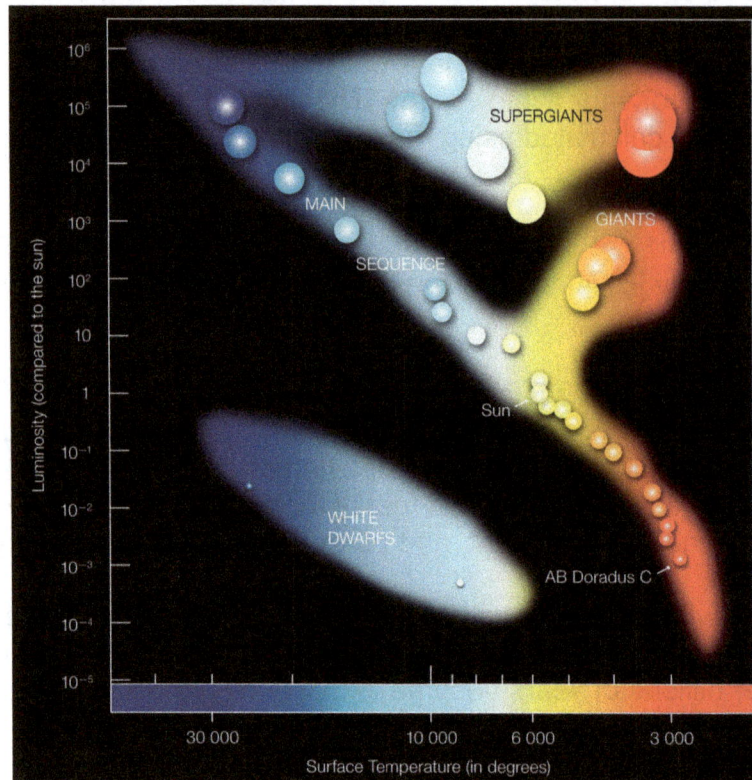

The Hertzsprung-Russell Diagram

Hertzsprung-Russell Diagram[viii]

Knowledge of the evolution of the Sun will enable us to make predictions for the future of Earth.

The Planets of Our Solar System[ix]

The Earth is one of eight planets in our Solar System. It is important to understand how the Earth was generated and the changes that have occurred to the Earth over time to enable future predictions. The presence of the other planets in our Solar System is important to the ability of the Earth to sustain life. For example, Jupiter often protects the Earth and the other inner planets by deflecting comets and asteroids.

As discussed previously, matter from the nebula collected in smaller clumps forming planetesimals. Most of the gas and small particles were swept from the inner Solar System, leaving only the proto-planets, moons, and asteroids. Further out smaller, colder bodies remained to fall inward as comets and meteors when they descended toward the planets and the Sun. After planets had attained most of their mass, intense meteor bombardment continued for approximately another half-billion years, and occasionally occurs during the following eons, even in our own epoch. Dating of lunar samples, meteorites, and impact sites on the inner planets provides data to model the timing of the meteor bombardment.

At the high temperatures of the inner nebula, the small proto planets (Mercury, Venus, Earth, Mars) were too hot to hold the volatile gases that dominated the nebula. Only high melting point materials like iron and rocky silicates were stable, so the terrestrial planets are made primarily of metallic cores and silicate mantles with atmospheres thin or absent. In the outer nebula, temperatures were cool enough for gases to accumulate and be held by proto planets. Jupiter, Saturn, Uranus, and Neptune are gas giants, made mostly from hydrogen, helium, and hydrogen compounds like ammonia and methane.

Planets are illuminated by closest stars. The atmosphere of planets can be determined by observation of the spectrum of light detected from the planets. As the light reflected from the planet passes through the atmosphere certain frequencies of light are filtered depending on the elements in the atmosphere. Thus, we can determine the constituents of the atmospheres of distant planets.

Initially the Earth was hot due to radioactive decay and from meteor impacts. The molten Earth allowed heavy elements such as iron to sink toward the center of the Earth forming the core of the planet. The iron also carried other elements that bind to it. Convection currents in the outer core are due to heat from the even hotter inner core. The heat that keeps the outer core from solidifying is produced by the breakdown of radioactive elements

17

in the inner core. Radioactive material in the Earth continues to decay. There is a finite amount of radioactive material in the core of the Earth that is used as fuel to heat the Earth. Knowledge of the rate of decay of this material can help us in predicting when this fuel will be depleted resulting in a cooling of the core, mantle, and crust of the Earth. We will revisit this issue in looking at distant future events.

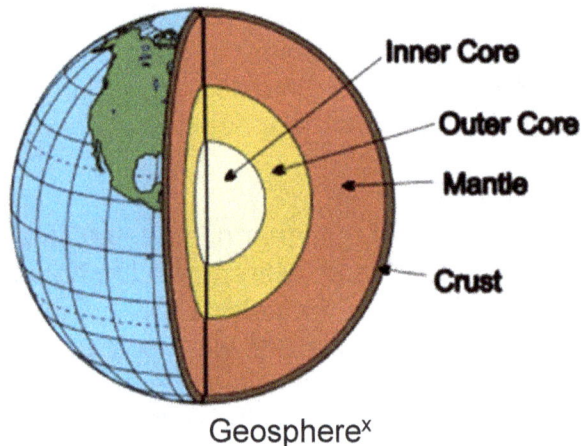

Geosphere[x]

Lighter material that is present in the crust moved to the surface of the planet forming our continents. Under the continents is a layer of solid rock known as the upper mantle. Though solid, this layer is ductile enough to slowly flow under heat convection, causing tectonic plates to move. Tectonic plates are massive, slabs of rock, generally varying from hundreds to thousands of kilometers across. Tectonic plate movement will be discussed in more detail later (see section on Tectonic Plate Movement).

Our Moon

There is evidence that the Moon formed from material from a collision of the Earth and planetesimal (small pre-planet) called Theia. Efforts to confirm that the impact had taken place centered on measuring the ratios between the isotopes of oxygen, titanium, silicon, and others[xi]. Computer models have allowed the mass of the incoming body, the young Earth, and the ejected material to be estimated. Most models estimate that the Moon is composed of around 70 to 90 percent material from Theia, with the remaining 10

18

to 30 percent coming from the early Earth. This impact is estimated to have occurred approximately 4.5 billion years ago based on analysis of rocks from the Earth and the Moon. The relatively strong protective magnetic field of the Earth from solar wind (continuous flow of charged particles from the Sun that permeates the Solar System) is in part due to the collision. The collision also resulted in a large moon that protects the Earth from most large meteor and comet impacts.

In 2000 geochemist Alex Halliday gave the incoming planetesimal the name Theia[xii], after the mother of the Moon goddess Selene in Greek mythology. This collision (see below) turned the newly formed Earth into a molten ball of rock again, and ejected material into orbit. Most of the material crashed back into the Earth, but some collected from mutual gravity to form the Moon.

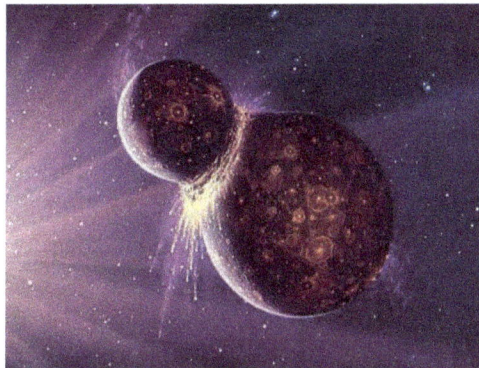

Artist's Impression of the Impact that Caused the Formation of the Moon: Credit NASA/GASFC[xiii]

The computer simulations predict that at the time of its formation, the Moon was closer to the Earth, 22,500 km away, compared with the 402,336 km between the Earth and the Moon today. The Moon continues to move away from the Earth, at the rate of 3.78 cm per year.[xiv] This movement away from the Earth would tend toward a slightly smaller tidal force[xv]. Thus, we can predict future decreases in the tidal force in the distant future unless this trend is somehow changed.

Land, Water, and Atmospheric Changes

The materials that accumulated into the early Earth were likely added over time. The early Earth was hot from gravitational collapse, impacts, and radioactive heating. Consequently, the early Earth was partially or largely molten. Volatile materials, carried by meteors and comets, were combined into the mantle by impacts that penetrated the interior. The early Earth was cooled largely by convection, which allowed hotter, less dense material, known as the crust, to rise and then transfer heat from the Earth.

Currently, about 2,900 kilometers below the Earth's crust, or surface, is the hottest part of our planet, called the core. Heat from the core is constantly conducting outward and warming rocks, water, gas, and other geological material. A small portion of the core's heat comes from the friction and gravitational pull formed when the Earth was created more than 4 billion years ago. The Earth would have cooled off and had a cold iron ball at the core if it were not for the continued release of heat by the decay of radioactive elements in the core. These radioactive elements include potassium-40, uranium-238 and thorium-232, which have half-lives of 1.25 billion, 4 billion and 14 billion years, respectively. Note: As mentioned previously, a half-life is time taken for the radioactivity of a specified isotope to become half of its original value. Thus, we can predict the rate of depletion of these elements and potential rate of cooling.

Radioactive decay is a continual process in the core. Temperatures there rise to more than 5000° C (about 9000° F), cooling as it spreads outward and heating the intervening mantle and crust above. If underground rock formations are heated to about 700-1300° C (1300-2400° F), they can become magma. Magma is molten (partly melted) rock permeated by gas and gas bubbles. Magma exists in the mantle and lower crust, and sometimes bubbles to the surface. When magma reaches the surface of the Earth it is called lava. Lava flows can often be seen at the outlets of volcanos.

Heat flow from the core maintains the convection of the metallic outer core and the geodynamo, which is the mechanism responsible for generation of the Earth's magnetic field. The outer core of the Earth is composed of an electrically conducting fluid (molten iron and nickel) whose motions produce a magnetic field. The exact direction and strength of the field fluctuates over time. The Earth's internal heat powers most geological processes and drives the movement of fractured sections of the crust called tectonic plates. The Earth's magnetic field protects the atmosphere from being stripped by the electrically charged solar wind, enabling life to persist.

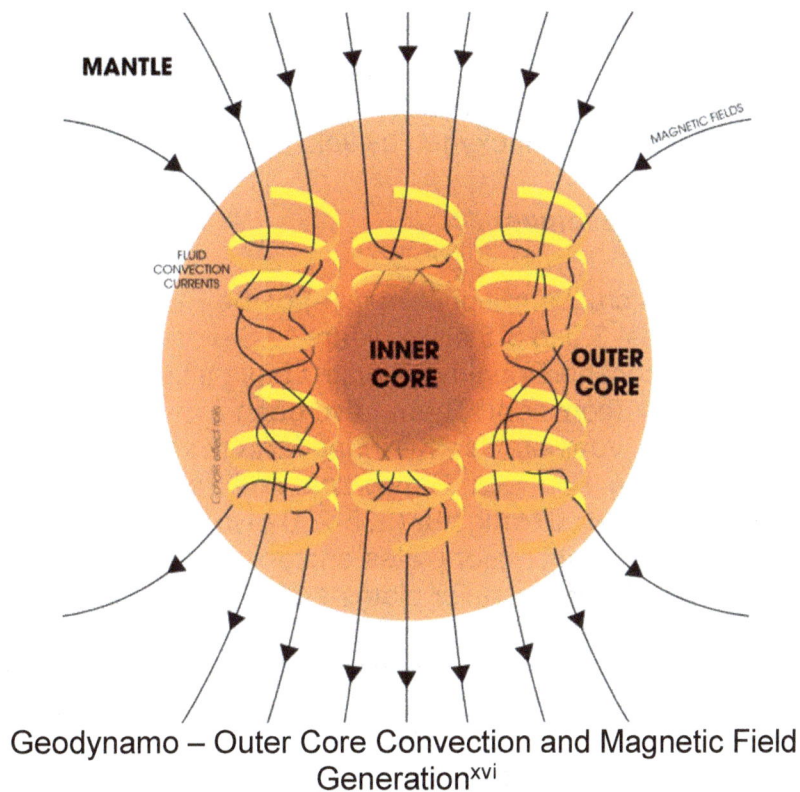

Geodynamo – Outer Core Convection and Magnetic Field Generation[xvi]

Early Earth seafloor spreading (the formation of new areas of oceanic crust or outermost layer of Earth's lithosphere that is found under the oceans) was likely not as well organized as today. The pattern of spreading and subduction (the sideways or downward movement of the edge of a tectonic plate of the Earth's crust into the mantle beneath another plate) was probably more vigorous and chaotic. There may have been rapid convective turnover of the mantle that resulted in rapid release of the volatiles (rapidly evaporating chemicals) stored in the mantle. Gases in the mantle escaped into the atmosphere when the magma erupted at

21

the surface. Some gasses, especially the lightest like hydrogen and helium, over eons evaporate into space.

Volcanic activity released large amounts of carbon dioxide, water vapor, and other gases into the atmosphere. Most of the released water vapor condensed to form the oceans. This outgassing formed the oceans and atmosphere. The addition of large amounts of oxygen to the atmosphere occurred later in history as a byproduct of life (i.e., marine organisms such as phytoplankton). Oxygen generating life forms have given rise over the past 3.5 billion years to our oxygen rich atmosphere, which is approximately 21 percent oxygen by volume.

Tectonic Plate Movement

The lithosphere (crust and upper mantle) of the Earth has been in motion for the last 3.3 to 3.5 billion years.[xvii] Several times in the past eons, the continents on these plates have all collided to form a worldwide supercontinent. There have been approximately ten supercontinents: Vaalbara, Ur, Kenorland, Arctica, Atlantica, Columbia, Rodinia, Pannotia, Gondwana, and Pangaea.[xviii] The most recent was called Pangaea (335-200 Mya).[xix] There are currently seven or eight major plates and dozens of smaller, or minor, plates[xx]. Six of the major plates are named for the continents embedded within them, such as the North American, African, and Antarctic plates. Though smaller in size, the minor plates are important when it comes to shaping the earth. The Juan de Fuca plate is largely responsible for the volcanoes in the Pacific Northwest of the United States (see below).

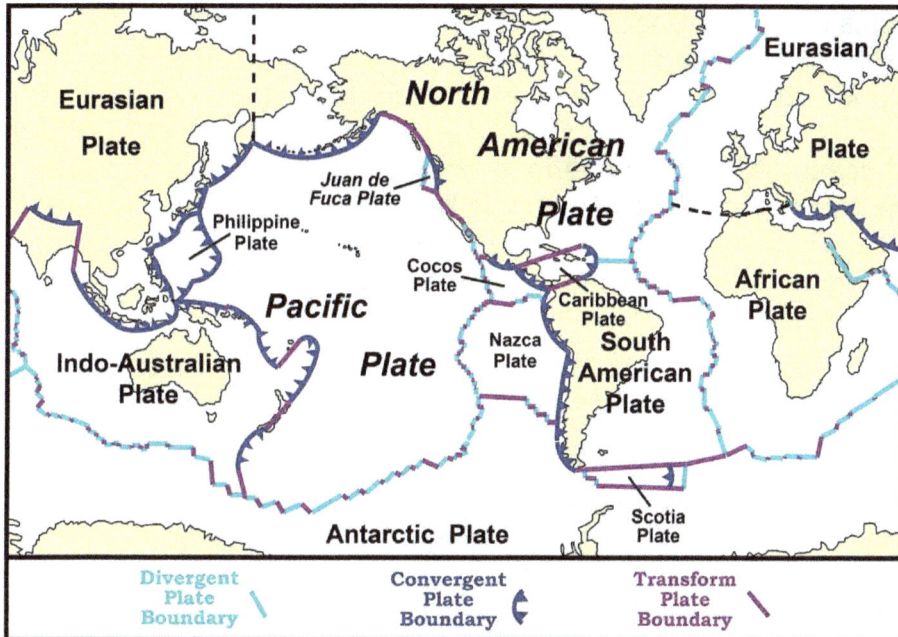

Plate Tectonics[xxi]

Convective currents in the molten rocks below the surface propel the tectonic plates. Most geologic activity including volcanos, earthquakes, and mountain formation stems from the interactions where the plates meet or divide.

The movement of the plates creates three types of tectonic boundaries:
1. Convergent - where plates move into each other
2. Divergent - where plates move apart
3. Transform - where plates move sideways in relation to each other.

The plates tend to move at a rate of 3-5 cm per year.

Where landmass plates collide, the crust buckles into mountain ranges. India and Asia collided about 55 Mya, slowly giving rise to the Himalayas, the highest mountain system on Earth. This process of collision is continuing today. These convergent boundaries also occur where a plate of ocean sinks, in a process called subduction, under a landmass. As the overlying plate lifts, it

23

forms mountain ranges. The sinking plate melts and is often expelled to the surface in volcanic eruptions such as those that formed some of the mountains in the Andes of South America.

At ocean-ocean convergences, one plate usually sinks beneath the other, forming deep trenches like the Mariana Trench in the North Pacific Ocean, the deepest point on Earth. These types of collisions can also lead to underwater volcanoes that eventually generate islands like Japan.

At divergent boundaries in the oceans, magma from deep in the Earth's mantle rises toward the surface and pushes apart two or more plates. Mountains and volcanos rise along the seam. The process renews the ocean floor and widens the giant basins. A single mid-ocean ridge system connects the world's oceans, making the ridge the longest mountain range in the world. These geologic tectonic processes help concentrate deposits of many rarer and useful elements such as metals and rare earth minerals. We will see that these elements have been and continue to be important in many of civilizations' technological advances.

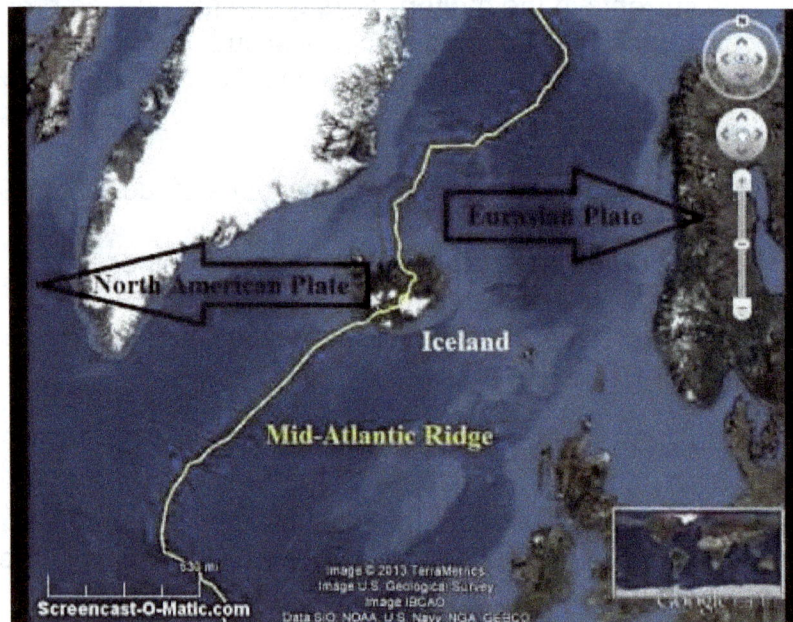

Mid-Ocean Ridge[xxii]

On land, giant troughs, such as the Great Rift Valley in Africa, formed where plates are tugged apart. If these plates continue to diverge, millions of years from now eastern Africa will split from the continent to form a new landmass. A mid-ocean ridge would then mark the boundary between the plates.

The San Andreas Fault in California is an example of a transform boundary, where two plates grind past each other along what are called strike-slip faults. These boundaries do not produce mountains or oceans, but the halting motion often triggers large earthquakes, such as the 1906 one that devastated San Francisco. The city of Los Angeles is on a plate moving in an approximately northward direction at about 5 mm a year[xxiii].

It is possible to watch a computer simulation that models tectonic plate movement for the last billion years, albeit the accuracy of such a model may be limited as the extrapolations are extended.[xxiv] Regardless, it provides an understanding of how mobile the plates are in geological time. It also provides insight into the various changes that have occurred and are continuing to occur, potentially allowing us to predict future events. This information provides insight into how the tectonic plate movements will continue to effect life on earth. Over long periods of time tectonic plate movement will change the size of and current flows of oceans, cause mountain formations, and relocate continents. Supercontinents over hotspots (small areas of the Earth's crust with an unusually high heat flow) can affect global heat circulation and sometimes vulcanism.

Climactic Changes

For thousands of years at a time, even the more temperate regions of the globe were covered with glaciers and ice sheets.

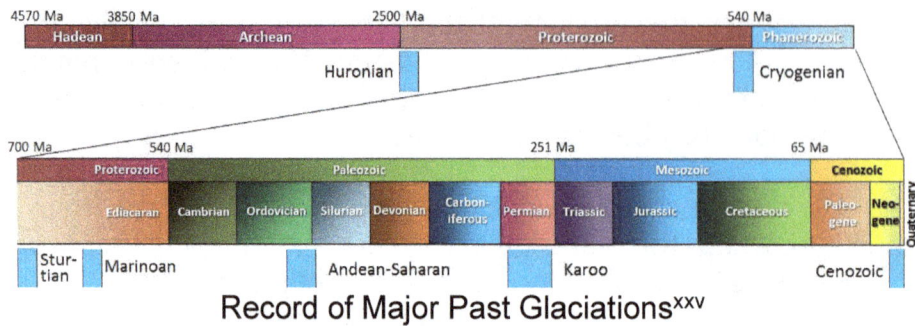

4570 Ma	3850 Ma		2500 Ma		540 Ma	
Hadean	Archean		Proterozoic			Phanerozoic

Huronian | Cryogenian

700 Ma	540 Ma				251 Ma		65 Ma	
Proterozoic		Paleozoic				Mesozoic		Cenozoic
	Ediacaran	Cambrian	Ordovician	Silurian	Devonian	Carbon-iferous	Permian	Triassic	Jurassic	Cretaceous	Paleo-gene	Neo-gene

Stur-tian | Marinoan | Andean-Saharan | Karoo | Cenozoic

Record of Major Past Glaciations[xxv]

The oldest known glacial period is the Huronian, which occurred 2.4 – 2.1 billion years ago. Late in the Proterozoic Period the Earth was in an intense glaciation. The Earth's surface may have been entirely or nearly entirely frozen, sometime earlier than 650 million years ago (Mya)[xxvi]. The glaciations of the Cryogenian Period are also known as the "Snowball Earth" glaciations, because the entire planet was frozen with ice on the oceans, hypothesized to be perhaps up to 1 km thick. The main theory as to what caused such an extreme global glaciation is a decrease in sunlight absorbed by the Earth's surface and incoming sunlight to the Earth. Potential causes include[xxvii]:

- enormous volcanic eruptions distributing sunlight-blocking particles into the stratosphere
- breaking up of the supercontinent (Rodinia) causing erosion of crustal rocks and consequently resulting in a reduction of carbon dioxide in the atmosphere, which produces a cooler climate
- Solar System encountering giant molecular clouds in the spiral arms of our galaxy reducing sunlight to the Earth

The Cryogenian Period is divided into two main glacial periods, Sturtian Period around 700 Mya and the Marinoan Period around 650 Mya.

The end of the Cryogenian Period glaciations coincides with the evolution of relatively large and complex life forms on Earth. This started during the Ediacaran Period and continued with the so-called explosion of life forms in the Cambrian Period (to be discussed in more detail later – see section on Origin of Life and Evolution). There have been three major glaciations during the

26

Phanerozoic Period (Andean-Saharan, Karoo, and the Cenozoic).

The Earth was warm and essentially unglaciated throughout the Mesozoic Period. The dinosaurs, which dominated terrestrial habitats during the Mesozoic, did not have to endure icy conditions.

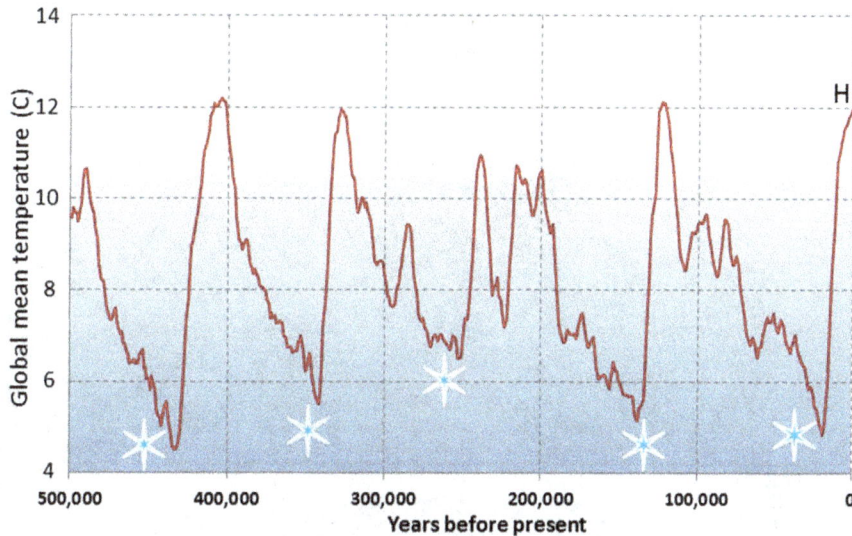

Ice Ages and Interglacial Periods Over the Last 500,000 Years[xxviii]

The most recent ice ages, which are better understood than the ancient snowball glaciations, occurred during the Pleistocene epoch, which lasted from 2.6 Mya to 11,700 years ago. This period has been characterized by significant temperature variations (through a range of almost 10°C) on time scales of 40,000 to 100,000 years, and corresponding expansion and contraction of ice sheets.

Over the past million years, the glaciation cycles have occurred approximately every 100,000 years. The last five glacial periods are marked with snowflakes (see above). The current interglacial (Holocene) is marked with an H. The most recent glacial period, which peaked at around 20 thousand years ago (kya), is known as the Wisconsin Glaciation. At the height of this last glaciation, massive ice sheets covered almost all of Canada and much of the northern United States

27

These periodic variations are attributed to Milankovitch cycles. Milankovitch cycles[xxix] describe how relatively slight changes in Earth's movement affect the planet's climate. These changes in the Earth's movements (eccentricity, obliquity, and precession) are explained in more detail below.

There is a link between the amounts of solar radiation in the high northern latitudes to previous European ice ages. There are three different positional cycles, each with its own cycle length, that influence the climate on Earth: the eccentricity of Earth's orbit, the planet's axial tilt, and the wobble of its axis (precession).

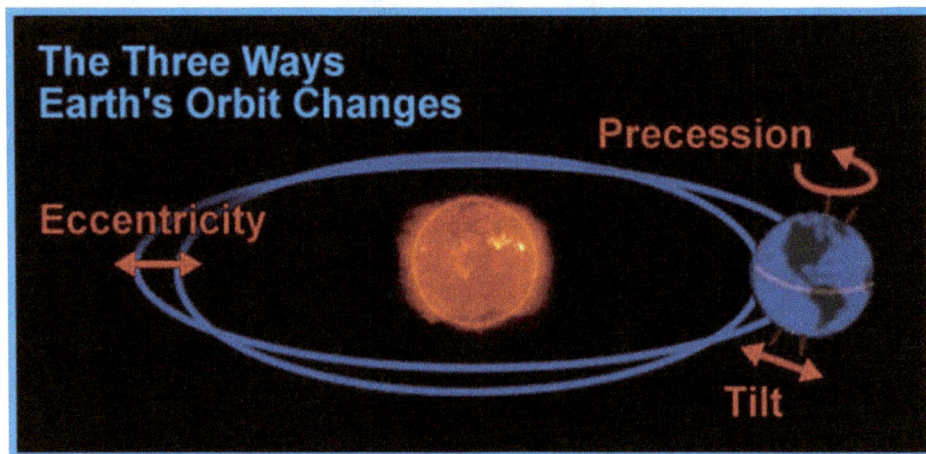

The Three Ways Earth's Orbit Changes[xxx]

The Earth orbits the Sun in an ellipse[2]. The Earth's orbit changes its eccentricity over the course of 100,000 years from nearly 0 to 0.07 and back again. When the Earth's orbit has a higher eccentricity, the planet's surface can receive 20 to 30 percent more solar radiation. Today, the eccentricity of the Earth's orbit is

_____

[2] Ellipticity is a measure of the shape of the oval and is defined by the ratio of the semi-minor axis (the length of the short axis of the ellipse) to the semi-major axis (the length of the long axis of the ellipse). A circle has an eccentricity of 0 and an ellipse that is almost a straight line has an eccentricity of nearly 1.

0.017, which produces about a 6 percent difference in solar radiation over the course of each year.

The tilt of the Earth's axis relative to the plane of its orbit is the reason that we experience seasons. Slight changes in the tilt changes the amount of solar radiation falling on certain locations of the Earth. Over the course of about 41,000 years, the tilt of the Earth's axis has varied between 21.5 and 24.5 degrees.

When the axis is at its minimal tilt, the amount of solar radiation does not change much between summer and winter for much of Earth's surface and therefore, seasons are less severe. This means that summer at the poles is cooler, which allows snow and ice to persist through summer and into winter, eventually building up into enormous ice sheets. Today, the Earth is tilted 23.5 degrees, and the tilt is slowly decreasing.

The Earth wobbles as it spins on its axis, similarly to when a spinning top begins to slow down. This wobble, known as precession, is primarily caused by the gravity of the Sun and Moon pulling on Earth's equatorial bulges[3]. The wobble does not change the tilt of Earth's axis, but the orientation changes. Over about 26,000 years, the Earth wobbles around in a complete circle, known as a precession cycle.

For the last several thousands of years, the Earth's axis has been pointed north toward Polaris. But the Earth's gradual change in wobble means that Polaris is not always directly north. About 5,000 years ago the Earth was pointed more toward another star, called Thubin. In approximately 12,000 years, the axis will have traveled a bit more around its precession circle and will point toward Vega, which will become the next North Star.

As the Earth completes a precession cycle, the orientation of the planet is altered with respect to perihelion (the point in the orbit of the Earth at which it is closest to the Sun) and aphelion (the point

---

[3] Equatorial bulges are an increase in the diameter of the Earth at the equator due to the centrifugal force exerted by the rotation about the Earth's axis.

in the orbit of the Earth at which it is furthest from the Sun). If a hemisphere is pointed toward the Sun during perihelion, it will be pointed away during aphelion, and the opposite is true for the other hemisphere. The hemisphere that is pointed toward the Sun during perihelion and away from the Sun during aphelion experiences more extreme seasonal contrasts than the other hemisphere. Currently, the southern hemisphere's summer occurs near perihelion and winter near aphelion, which means the southern hemisphere experiences more extreme seasons than the northern hemisphere.

Thus, we can see that the Earth's climate changes are often cyclical and are affected by several different factors with differing periodicity.

Origin of Life and Evolution

Life on Earth likely started as chemical reactions in areas with heat and water, such as in volcanic vents in the ocean. Life requires an energy source, which can be provided by the Sun or by geothermal sources. The changes and increasing complexity of life are a product of evolution. Changes in genetic information are needed for evolution. Evolution is counter to entropy (tendency toward disorder or more probable states) but has occurred due to the localized increase in energy in an open system such as the Earth. Note: Entropy will increase in a closed system (a system that does not allow matter or energy from outside environments to enter its space). While the details of how the first lifeform originated are not completely understood, fossilized microorganisms have been found in rocks as old as 3.4 billion years[xxxi]. The Miller-Urey experiment[xxxii] provides some insight into how amino acids (possible building blocks for lifeforms) many have formed from smaller molecules such as water, hydrogen, methane, and ammonia.

Many of the same simple organic molecules have been found in meteorites, comets, and interstellar gas clouds and could have been sown by meteorites or comets colliding with the Earth[xxxiii]. The simplest of the molecules found in living systems seem to be common in nature. In the 1980s, Thomas Cech and Sidney Altman showed that some ribonucleic acid (RNA) molecules can act as enzyme-like catalysts for more complex reactions[xxxiv]. Something like RNA was likely assembled and was then able to fill the roles of an enzyme and a hereditary molecule in a precursor to life. The RNA systems were then acted upon by natural selection, leading to greater molecular complexity containing and replicating information through successive generations. Cech and Altman's work demonstrate that life went through an early RNA-dominated phase. RNA can encode genetic information in its structure for use in reproduction and function.

A primordial molecule provided both catalytic power and the ability to propagate its chemical identity over generations. As the catalytic versatility of these primordial RNA molecules increased due to random variation and selection, metabolic complexity began to emerge. It is noteworthy that RNA still catalyzes several fundamental reactions in modern-day cells, which can be viewed as molecular fossils. Cells are compartments in which biochemical reactions are sped and interlinked more easily in a solution of water. Cell walls, if repellent of water, can form to enclose the water and biomolecules.

RNA and the Origin of Life

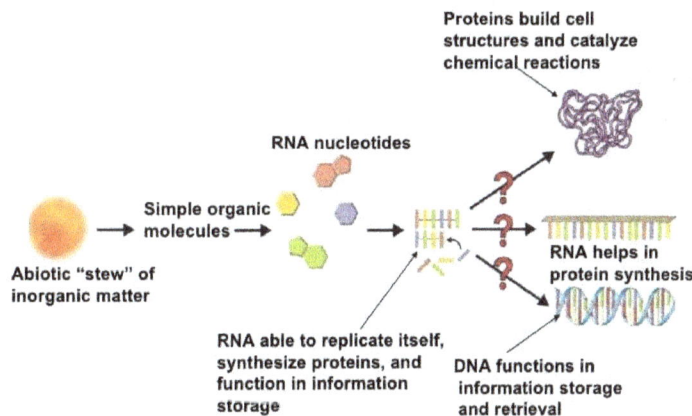

RNA and the Origin of Life[xxxv]

The earliest reactions could have occurred in porous rock, perhaps filled with organic gels. This early version of metabolism likely

31

consisted of a series of simple chemical reactions running without the aid of complex enzymes, through the catalytic action of networks of small molecules, perhaps aided by naturally occurring minerals.

Almost a billion years later, sunlight was used in a photochemical process known as photosynthesis to extract molecular energy to sustain life. There are two types of photosynthetic processes: oxygenic (most common and generates oxygen) and anoxygenic photosynthesis (does not generate oxygen, typically occurs in bacteria). Oxygenic photosynthesis is the process to turn sunlight, carbon dioxide, and water into glucose and oxygen. Oxygenic photosynthesis incorporates a combination of carbon, hydrogen, and glucose into the lifeform (currently plants, algae, some protists, or cyanobacteria) from carbon dioxide and water, releasing oxygen as a waste product.

Animals use the by-product of plants, oxygen, for respiration, which in turn generates carbon dioxide needed by the plants; thus forming a carbon cycle between plants and animals. Fungi such as mushrooms further recycle the carbon, hydrogen, and trace elements.

As compared to RNA, deoxyribonucleic acid (DNA) is a more evolved molecule that not only stores information but preserves it by its redundancy in a double stranded helix structure. DNA and RNA are both nucleic acids. Alongside proteins, lipids and complex carbohydrates, nucleic acids are one of the four major types of macromolecules that are essential for all known forms of life. The two DNA strands are known as polynucleotides as they are composed of simpler monomeric units called nucleotides. Each nucleotide is composed of one of four nucleobases (cytosine, guanine, adenine or thymine), a sugar called deoxyribose, and a phosphate group.

Earliest life may have formed between 3.5 and 4 billion years ago during Archean Era. These prokaryotes have neither a distinct nucleus with a membrane nor other specialized organelles. Prokaryotes include the bacteria, archaebacteria (ancient intermediate group between the bacteria and eukaryotes), and cyanobacteria. Some archaebacteria persist to present. There are microbial mat fossils such as stromatolites (layered forms of sedimentary rocks that are created by cyanobacteria) found in 3.48 billion-year old sandstone discovered in Western Australia. Photosynthetic organisms appeared between 3.2 and 2.4 billion years ago and began enriching the atmosphere with oxygen.

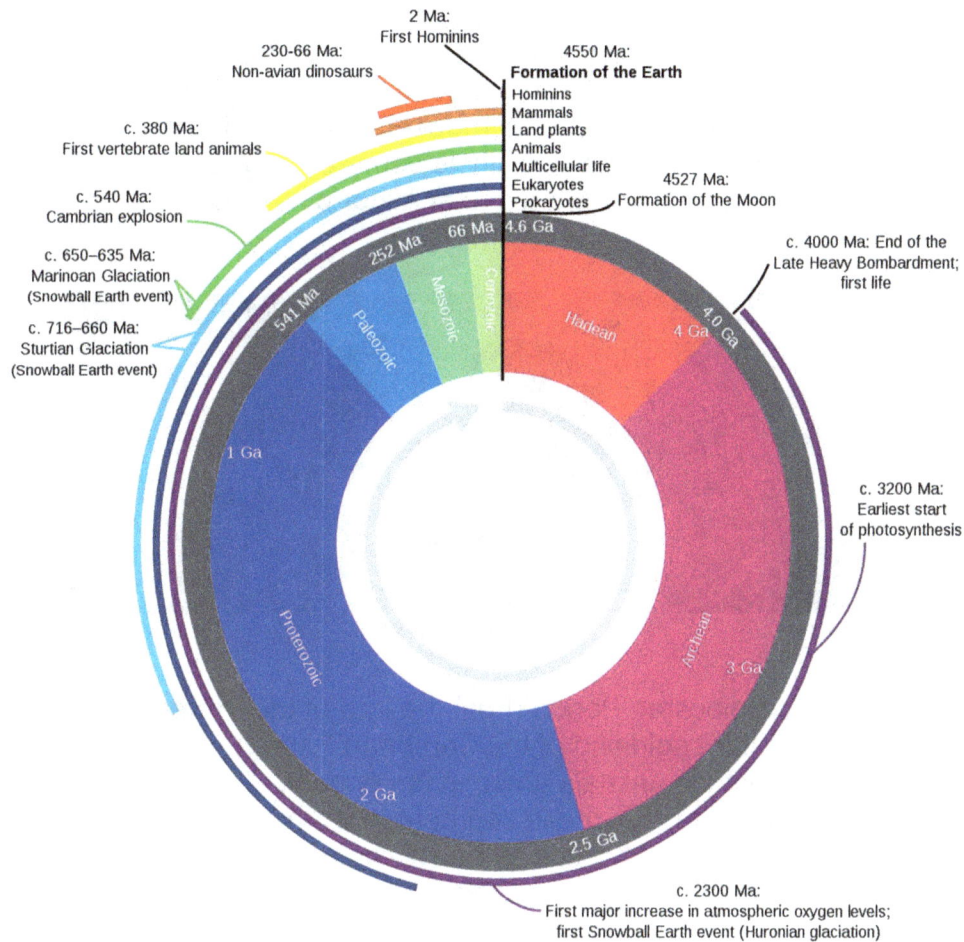

History of Life on Earth[xxxvi]
Note: Ma is million years ago, Mya is used elsewhere in the book, and Ga is billion years ago.

During the Proterozoic Eon, eukaryotes (organisms with a cell in which the genetic material is contained within a distinct nucleus) and multi-cellular life emerged. Protists (organisms having a fully

33

defined nucleus, with complex cellular structure) developed approximately 1.8 billion years ago. Some examples of protists include amoebas, diatoms, and paramecium.

Late in the Proterozoic Era the first animals arose, perhaps as early as 800 Mya. Sponges and jellies were among the earliest animals. Oxygen levels in the ocean were low, but sponges can tolerate conditions of low oxygen. Later multicellular organisms such as segmented worms, fronds, disks, or immobile bags evolved during the Ediacaran Era (635 Mya to 539 Mya). Charnia consisted of a leaf-like frond, composed of alternating primary branches, each made-up of multiple semi-rectangular secondary modular elements, tethered to the seafloor by a basal attachment bulb.

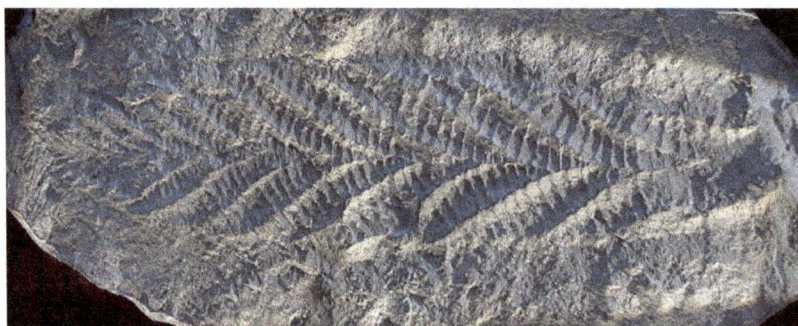

Charnia[xxxvii]

During the Ediacaran Period (from the end of the Proterozoic Eon 635 Mya to the beginning of the Cambrian Period), simple but varied soft-bodied animals arose in the oceans. The most common types of organisms resemble segmented worms, fronds, disks, or immobile bags. Motility such as burrowing and crawling appears to have developed during this period.

The Cambrian Explosion of many more advanced body designs occurred around 542-488 Mya. The development of hard body parts such as shells, skeletons or exoskeletons in animals like mollusks, echinoderms, chordates, crustaceans, bivalves, crinoids, chordates, and arthropods (a well-known group of arthropods from the lower Paleozoic are the trilobites) made the preservation and fossilization of such life forms easier than those of their Ediacaran and Proterozoic ancestors. For this reason, much more is known

34

about life in and after the Cambrian Period than about that of older periods, such as the Ediacaran, where trace fossils predominate.

The first land plants appeared around 470 Mya, during the Ordovician Period, when life was diversifying rapidly. They were non-vascular plants, like mosses and liverworts, that did not have deep roots. Invertebrate life became more diverse and complex through the Ordovician Period. Both calcareous and siliceous sponges are known; including the stromatoporoids that first appeared in the Ordovician Period. Echinoderms (invertebrates such as starfish, sea urchins, or sea cucumber) reached their peak diversity, with crinoids (a group related to present-day sea lilies and feather stars), cystoids (spherical, stalked echinoderms), asteroids (starfishes), edrioasteroids (barnacle-like), and homalozoans (asymmetrical echinoderms) being the most common. Fossils of primitive fish have also been identified in this period.

Amphibians and reptiles followed. The first major groups of amphibians developed in the Devonian period, around 370 Mya from fish with lungs and bony-limbed fins, which were helpful in adapting to dry land. The first reptiles arose about 310 Mya during the Carboniferous period.

The first mammals evolved from a population of vertebrates called therapsids (mammal-like reptiles) at the end of the Triassic Period (around 200 Mya) and coexisted with dinosaurs throughout the Mesozoic Era. The first mammals were small and capable of generating heat. They were probably nocturnal to escape predation.

Evidence of the first hominins (species regarded as human) has been found in the Great Rift Valley of East Africa. During the later Miocene (10 – 5.5 Mya) and the early Pliocene (5.5 – 4 Mya) at least one lineage of apes made the transition to terrestrial and upright bipedal capability[xxxviii]. This shift may have been brought about by climate changes that were occurring in equatorial Africa at the time due to the geological rifting in the area caused by movement of crustal tectonic plates in East Africa. Gorillas, chimpanzees, bonobos, and orangutangs, and apes evolved in the millions of following years.

While we recognize humans as having consciousness or awareness of internal and external existence, it is unclear which life forms first evolved to have consciousness. The challenge of determining this is partly due to the problem with a clearly accepted definition of consciousness and with measuring this property or capability. While it appears reasonable that many or all animals have consciousness, it is unclear whether many or all plants, bacteria, etc. have consciousness.

Extinctions

There have been major extinctions due to environmental and other circumstances. Some of the biggest causes of mass extinctions include:

- Ocean/atmosphere chemistry
- Climate change
- Volcanic activity
- Meteor/asteroid impacts
- Formation and separation of supercontinents

There are a variety of causes for extinctions. Some speculate that an asteroid may again impact the Earth causing the devastation of life on Earth. Small objects frequently collide with the Earth. There is an inverse relationship between the size of the object and the frequency of such events. The lunar cratering record shows that the frequency of impacts decreases as approximately the cube of the resulting crater's diameter (frequency is inversely proportional to the crater diameter to the third power), which is on average proportional to the diameter of the impactor. Asteroids with a 1 km diameter strike Earth every 500,000 years on average[xxxix]. Large collisions, with 5 km objects, happen approximately once every 20 million years. The last known impact of an object of 10 km or more in diameter was at the Cretaceous–Paleogene extinction event 66 Mya.

Impact conditions such as asteroid size and speed, but also density and impact angle determine the kinetic energy released in an impact event. The more energy that is released, the more damage is likely to occur on the ground due to the environmental effects triggered by the impact. Such effects can be shock waves, heat radiation, the formation of craters with associated earthquakes, and tsunamis (if a water body is hit). Animal populations and plant life are vulnerable to these effects if they exist within the affected zone. It is possible that future asteroids could impact the Earth, which could lead to significant environment changes. The extent of these changes would determine the changes in species extinctions.

The movement of tectonic plates can cause changes in ocean currents and changes in vulcanism potentially producing vast volcanic events. Volcanoes can form when tectonic plates collide, and one plate is pushed beneath another. Hot magma rises from the mantle at mid-ocean ridges pushing the plates apart. The magma then rises and erupts as lava on the surface.

The formation and break-up of the supercontinent of Pangea can provide an example of the effect of plate tectonics on climate[xl]. Pangea was completely surrounded by a world ocean (Panthalassa) extending from pole to pole and spanning 80 percent of the circumference of Earth at the equator. The equatorial current system, driven by the trade winds, resided in warm latitudes much longer than today, and its waters were therefore warmer. As the dispersal of continents following the breakup of Pangea continued, the surface circulation of the oceans began to approach the more complex circulation patterns of today. About 100 Mya, the northward drift of Australia and South America created a seaway around Antarctica, which remained centered on the South Pole. A vigorous circum-Antarctic current developed, isolating the southern continent from the warmer waters to the north. The equatorial current system became blocked, first in the Indo-Pacific region and next in the Middle East and eastern Mediterranean about 6 Mya, by the emergence of the Isthmus of Panama. The equatorial waters were heated less, and the midlatitude ocean gyres (large rotating oceans currents) were not as effective in keeping the high-latitude waters warm. Thus, an ice cap began to form on Antarctica some 20 Mya and grew to roughly its present size about 5 million years later.

Displacement or destruction of some species by introduction of other successful species can cause significant changes to the environment (i.e., atmospheric changes, water pH changes, global warming) and can lead to the extinction of some species. Over hunting (i.e., Woolly Mammoth, Caspian Tigers, Dodo) can lead to extinction of the vulnerable species. This appears to be occurring due to the introduction of humans. This will be discussed further later (see section on Sixth Extinction – Holocene Extinction).

Some of these past causal agents for extinctions are likely to occur again in the future.

Some of the largest past mass extinctions include the Ordovician–Silurian, Late Devonian, Permian–Triassic, Triassic–Jurassic, and the Cretaceous-Paleogene Extinctions.[xli]

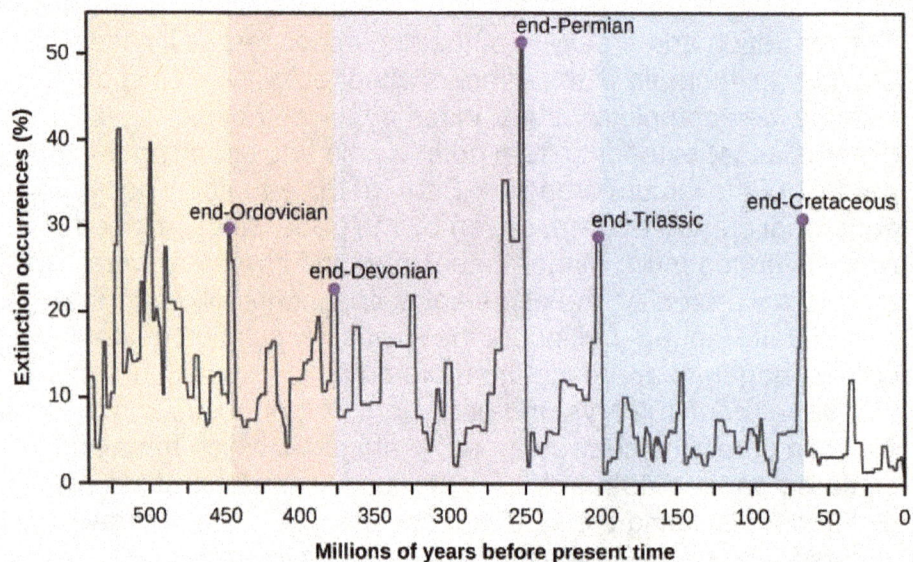

Major Extinctions Over Time[xlii]

Ordovician–Silurian Extinction (Late Ordovician Extinction)

The Ordovician–Silurian Extinction consists of two consecutive, closely timed, mass extinctions. The first is typically attributed to the Late Ordovician glaciation, which expanded over Gondwana and shifted the Earth from a greenhouse to icehouse climate. Gondwana was a supercontinent that existed from the Neoproterozoic (about 550 Mya) and began to break up during the Jurassic Period (about 180 Mya). As the southern supercontinent Gondwana drifted over the South Pole, ice caps may have formed on it.

The second extinction of the end-Ordovician occurred as the glaciation receded and warm conditions returned, associated with oxygen depletion and sulfide production both toxic to oceanic lifeforms. When combined, this change in ocean chemistry is widely considered to be the second most catastrophic extinction event in history.

Continents During Early Ordovician[xliii]

About 450–440 Mya, 60% to 70% of all species became extinct. This included 85% of marine species that died. The extinction caused the disappearance of one third of all brachiopod and bryozoan families, as well as numerous groups of conodonts, trilobites, echinoderms, corals, bivalves, and graptolites. Fishes survived and diversified following the mass species extinctions, as well as the horseshoe "crab" arthropod still surviving today in the oceans.

Ordovician Period - Trilobite[xliv]

The primary cause of the Ordovician–Silurian Extinction is believed to be due to the massive glaciation as mentioned and a sea level drop. As continental drift carried Gondwana to the South Pole, this locked water into ice caps. Overall, sea levels substantially dropped. Another possible causal agent includes toxic metals unleashed by depleted oxygen in the oceans[xlv]. The toxic metals may have killed life forms in lower levels of the food chain, causing a decline in population, and subsequently resulting in starvation for the dependent higher feeding life forms in the chain.

Late Devonian Extinction

The Late Devonian Extinction was less severe than the other mass extinctions, but overall, 19% of all families and 50% of all genera became extinct. This extinction occurred between 375–360 Mya. Some consider the extinction to be as many as seven distinct events, spread over about 25 million years. Below is a map showing the locations of continents during the Devonian period.

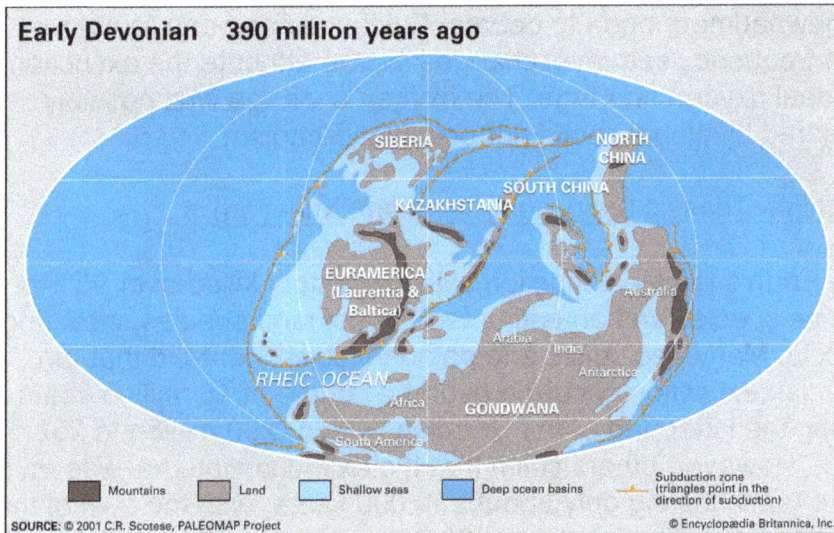

Early Devonian   390 million years ago

SIBERIA
NORTH CHINA
SOUTH CHINA
KAZAKHSTANIA
EURAMERICA (Laurentia & Baltica)
Australia
Arabia
India
Antarctica
RHEIC OCEAN
Africa
GONDWANA
South America

| Mountains | Land | Shallow seas | Deep ocean basins | Subduction zone (triangles point in the direction of subduction) |

SOURCE: © 2001 C.R. Scotese, PALEOMAP Project   © Encyclopædia Britannica, Inc.

Continents During the Early Devonian (390 Mya)[xlvi]

The extinction seems to have only affected marine life. Extinctions included some brachiopods, trilobites, and reef-building organisms. Reef-building organisms almost completely disappeared. Both graptolites and cystoids disappeared during this event.

Fig. 37.   Rekonstruktion einer Grapho-lithen-Qualle.

Graptolite[xlvii]

It is difficult to assign a single cause, and indeed to separate cause from effect. The sedimentary record shows that the late Devonian was a time of environmental change, directly affecting organisms and causing extinction. What caused these changes is

41

somewhat more open to debate. Evidence has been found for high-frequency sea-level changes. Some attribute the extinction to a global oxygen shortage. The oxygen shortage was possibly triggered by global cooling or oceanic volcanism.

Permian–Triassic Extinction (Permian Extinction)

The Earth's largest extinction event in history killed over 95% of all marine species and an estimated 70% of land species, including insects. Many geologists and paleontologists contend that the Permian extinction occurred over the course of 15 million years during the latter part of the Permian Period (299 million to 252 Mya). However, others claim that the extinction interval was much more rapid, lasting only about 200,000 years, with the bulk of the species loss occurring over a 20,000-year span near the end of the period.

Shallow warm-water marine invertebrates, which included the trilobites, rugose (wrinkled or corrugated) and tabulate corals, and two large groups of echinoderms (blastoids and crinoids), show the most-protracted and greatest losses during the Permian extinction.

Fossil of Crinoid[xlviii]

Several groups of aquatic vertebrates, such as the acanthodians, thought to be the earliest jawed fishes, and the placoderms, a group of jawed fishes with significant armor, were also eliminated.

42

Notable terrestrial groups, such as the pelycosaurs (fin-backed reptiles), moschops (a massive mammal-like reptile), and numerous families of insects also became extinct. In addition, a number of other groups, such as sharks, bony fishes, brachiopods, bryozoans, ammonoids, therapsids, reptiles, and amphibians experienced significant declines by the end of the Permian Period.

30 cm
12 inches

Dimetrodon

© 2015 Encyclopædia Britannica, Inc.

Pelycosaur[xlix]

It is difficult to define the underlying cause of the Permian-Triassic Extinction, but its coincidence with the collection of continents into one supercontinent during this time points to this geological change as an important underlying cause of mass extinction. Desertification on land and reduction of shallow seas at the continental margins would have reduced the capacity of the Earth to host some regional ecosystems. Hypotheses of temperature crises, especially of those occurring in shallow marine (surface) waters, are based in part on studies of oxygen isotopes and the ratios of calcium to magnesium in Permian fossil shell materials. The highest estimated temperatures of the ocean surface waters were estimated to be 25–28 °C (about 77–82 °F). Several studies have suggested that changes in the carbon isotope record may indicate a disrupted biological cycle. Some scientists consider the unusually high amounts of carbon-12 trapped in Permian sediments to be a result of widespread oceanic anoxia (low levels of dissolved oxygen). They associate this anoxia with the prolonged eruption of the Siberian flood basalts (called Siberian Traps), which probably led to higher levels of carbon dioxide in the atmosphere. The rapid blooming of archaea (single-celled prokaryotic organisms), which evolved the ability to manufacture methane near the end of the Permian Period, may have played a significant role in the rise in Earth's ocean temperatures and changes to the planet's carbon cycle. A sudden increase in methane in the atmosphere is thought to result in warming temperatures, ocean acidification, and other changes to the carbon cycle. Note: We will discuss later similar circumstances in changes to the atmosphere currently occurring that also is leading to ocean acidification (see section on Oceanic Disruption).

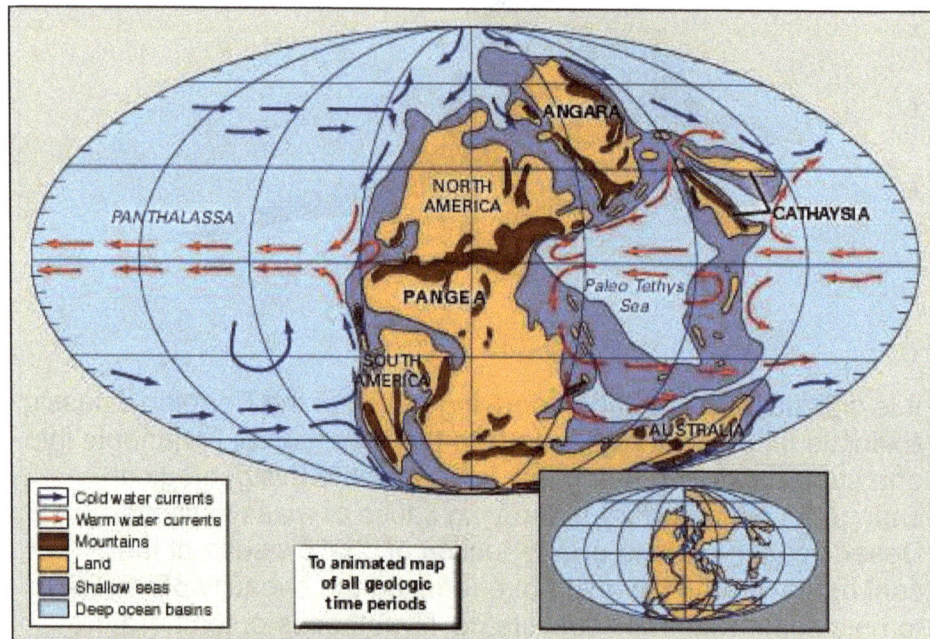

Continents and Current Flow during the Permian Period[I]

Triassic–Jurassic Extinction

The Triassic–Jurassic Extinction occurred 201.3 Mya, which marked the transition between the Triassic and Jurassic periods. This extinction witnessed approximately 70-75% of all terrestrial and marine species go extinct. This occurred before Pangea started to break apart. In the seas, the entire class of conodonts (jawless vertebrates resembling eels) and 23–34% of marine genera disappeared.

Conodonts[li]

On land, all archosauromorphs (reptiles similar to lizards) other than crocodylomorphs (crocodilian reptiles), and pterosaurs (flying reptiles) went extinct; some of the groups that died out were previously abundant, such as aetosaurs (armored herbivores), phytosaurs (semiaquatic reptiles), and rauisuchids (big-headed reptiles). What was left fairly untouched were plants, dinosaurs, pterosaurs, and mammals; this allowed the dinosaurs and pterosaurs to become the dominant land animals for the next 135 million years.

Desmatosuchus spurensis
50 cm

Aetosaur[lii]

Possible causes of extinctions include volcanic eruptions and giant flood basalts in the Central Atlantic Magmatic Province (CAMP), sea-level fluctuations, or oceanic acidification. The sudden release of carbon dioxide may have had caused global warming and amplified the greenhouse effect. The record of CAMP degassing shows several distinct pulses of carbon dioxide immediately following each major pulse of magmatism, at least two of which

45

amount to a doubling of atmospheric carbon dioxide[liii]. In the diagram below, the solid spots identify areas with existing basalt formed during volcanic eruptions resulting in lava flows.

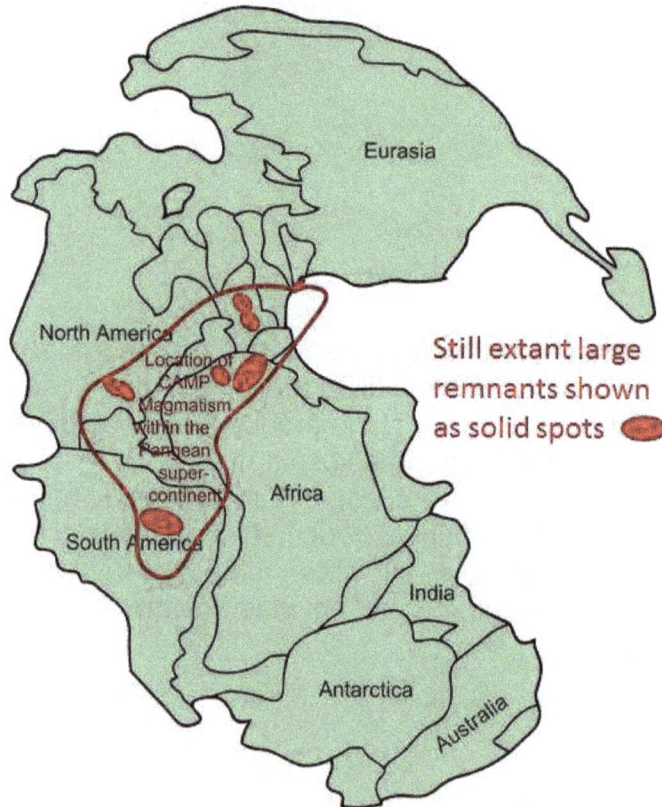

Maximum Extent of CAMP Volcanism at the Triassic-Jurassic Transition[liv]

Cretaceous-Paleogene Extinction

Approximately 66 Mya, about three quarters of all species became extinct during the transition from the Cretaceous to the Paleogene period[lv]. This is the period of the extinction of the dinosaurs. Fossils of bone, teeth, trackways, and other evidence demonstrate that the dinosaurs existed for at least 230 million years before this extinction.

In the oceans, ammonites (marine mollusk animals) disappeared. Other sea reptiles such as mosasaurs (large marine reptiles), ichthyosaurs (aquatic reptiles with porpoise-like heads), and plesiosaurs (marine reptile with a broad flat body and paddle-like limbs) became extinct earlier in the Cretaceous period. Lines of archosaurs, the group of reptiles that contains the dinosaurs, birds, and crocodilians, that survived the extinction were the lineages that led to modern birds and crocodilians. There is evidence of widespread extinctions of angiosperms (seed producing, flowering plants) and other dramatic shifts among North American plant communities. Eventually, mammals emerged as the dominant large land animals.

The cause of this extinction event is believed to be from an asteroid impact that left a 100-mile diameter (160 km) crater called the Chicxulub Crater, located along the coast of Mexico's Yucatan Peninsula[lvi]. The asteroid impact likely hurled gas, dust, and debris (from both the asteroid and the ground) into the atmosphere that altered the climate by blocking sunlight for many years, leading to deaths of plants from lack of energy and deaths of animals from the lack of food and from freezing as sunshine was diminished worldwide.

Dinosaur Extinction Asteroid[lvii]

There is also evidence of ancient lava flow in India known as the Deccan Traps with outpourings of lava between 60-65 Mya. These volcanic coverings are nearly 200,000 square miles (520,000 square kilometers) in layers that are in places more than 6,000 feet (1800 meters) thick. These lava flows would have deposited carbon dioxide, sulfur dioxide, and other gases into the atmosphere that contributed to cooling the climate. It may be that both events contributed to the Cretaceous-Paleogene extinction.

Deccan Traps in India[lviii]

Knowledge of previous extinctions and their prevalence provides insight into the potential future of life on Earth. It seems likely that additional periods of extinction will occur. Though no evidence yet exists from past extinctions, passage of the solar system through dense clouds in the galaxy or passing of another star in the outer solar system could alter the amount of light striking the surface of the Earth, leading to mass extinctions. Regardless, it appears that we today are now in a period of extinction that is sometimes referred to as the Holocene Extinction.

Sixth Extinction – Holocene Extinction

According to a 1998 survey of 400 biologists conducted by New York's American Museum of Natural History, nearly 70% believed that the Earth is currently in the early stages of a human-caused mass extinction, known as the Holocene extinction (sometimes referred to as the Anthropocene extinction, due to the impact of humans). In that survey, the same proportion of respondents agreed with the prediction that up to 20% of all living populations could become extinct by 2028. This extinction that is due to the rapid proliferation of humans, and the consequential actions of humans, is causing a rapid extinction of many species. This will be discussed in more detail later as we cover global warming and changes to our environment (see section on Global Warming).

Humanoids and Humans

Human evolution is the process of change by which people originated from apelike ancestors. Humans and the great apes of Africa, chimpanzees (including bonobos) and gorillas, are believed to have shared a common ancestor that lived between 8 and 6 Mya. Humanoids (a being resembling a human in it shape) likely appeared around 6 Mya as evidenced by hominid (group consisting of all modern and extinct great apes) fossils.[lix] It appears that humans first evolved in Africa, and much of human evolution occurred on that continent. The fossils of early hominids who lived between 6 and 2 Mya come from Africa[lx]. Some hominoid species from this period exhibit traits that are typical of humans but are not seen in the other living apes, leading paleoanthropologists to infer that these fossils represent early members of the hominin lineage. The first human-like traits to appear in the hominin (a primate those species are regarded as human, ancestral to humans, closely related to humans) fossil record are bipedal walking and smaller, blunt canines. African *Homo erectus* or *Homo ergaster* may have been the first human species to leave Africa. Fossil remains show this species had expanded its range into southern Eurasia by 1.75 Mya[lxi]. Their descendants, Asian *Homo erectus*, then spread eastward and were established in Southern East Asia by at least 1.6 Mya.

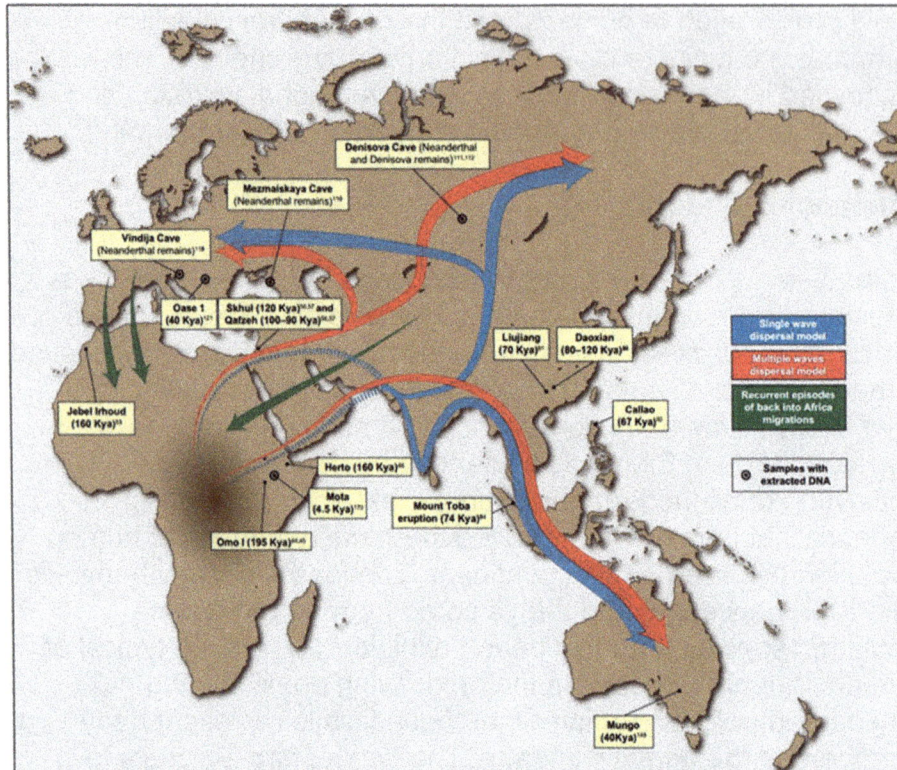

Migration Waves Out of Africa (Kya = thousand years ago)[lxii]

From both fossils and molecular clocks (a method that uses the mutation rate of biomolecules to deduce the time in prehistory when two or more life forms diverged), modern *Homo Sapiens* evolved in Africa between 300,000-225,000 years ago[lxiii]. Afterwards, they expanded into Asia, Australia, and Europe, arriving in the Americas approximately 15,000-20,000[lxiv] years ago but with subsequent more recent migrations from Asia.

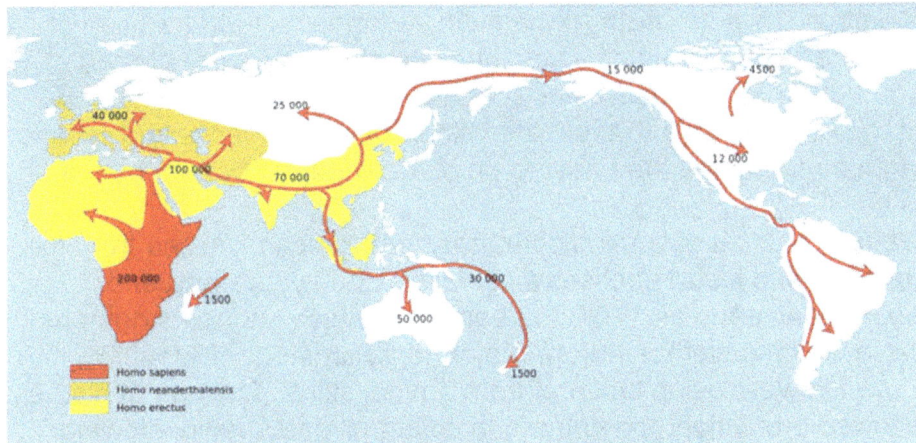

The Spreading of *Homo Sapiens*[lxv]

A supervolcanic eruption occurred around 75,000 years ago at the site of present-day Lake Toba in Sumatra, Indonesia. The Youngest Toba eruption is one of the Earth's largest known explosive eruptions. This event may have caused a global volcanic winter (volcanic ash, droplets of sulfuric acid, and water increasing reflection of solar radiation causing cooling) of six to ten years and possibly a 1,000-year-long cooling episode. This eruption has been linked to a genetic bottleneck in human evolution about 70,000 years ago, which may have resulted in a severe reduction in the size of the total human population due to the effects of the eruption on the global climate. According to the genetic bottleneck theory, between 50,000 and 100,000 years ago, human populations decreased to 3,000–10,000 surviving individuals. It is supported by some genetic evidence suggesting that today's humans are descended from a population of between 1,000 and 10,000 breeding pairs that existed about 70,000 years ago.

Changes in Civilization

Early humanoids learned to use tools as early as 2.6 Mya by making basic stone implements. By about 1.76 Mya humans began to make hand axes and cutting tools. Human language was unlikely to have begun until at least 200,000 years ago, as evidenced by fossils showing a lower placement of the larynx[lxvi] and hyoid[lxvii], which is needed for effective speech. Language aided in organized hunting. Analyses of decorated red ochre from

51

Blombos Cave in South Africa indicate that a cultural tradition of creating meaningful geometric designs stretched from around 100,000 to 75,000 years ago. Organized hunting using weapons made humans able to compete with larger, stronger animals. Oral language enabled the passing on of knowledge and information.

Genetic studies of lice indicate that clothing lice diverged from their human head louse ancestors at least 83,000 years ago and possibly as early as 170,000 years ago, which suggests humans were wearing clothes before major migrations out of Africa[lxviii]. Humans were using clothing from animal skins. The use of clothes allowed protection and warmth in colder climates, also allowing expansion.

The use of fire is evidenced as early as 40,000 years ago by the presence of flints used in starting fires. Fire enabled an increase in caloric intake. Cooking meat makes it easier to chew and unwinds proteins within it, allowing for better digestion. Fire also breaks down indigestible cellulose and starch in plants that would otherwise be inedible. Remains of art and musical instruments from this time show that human culture and probably brain function continued evolving during this period.

Around 30,000 years ago, every species of primate including *Homo neandertalis* appears to have vanished except for *Homo sapiens*[lxix], due partly to interbreeding, competition, and possibly violence.

Farming began around 12,000 years ago, which enabled hunter gathers to obtain a reliable food supply. The wild progenitors of crops including wheat, barley, and peas are traced to the Near East region. Cereals were grown in Syria around 9,000 years ago, while figs were cultivated even earlier. The origins of rice and millet farming date to around 6,000 BCE. Cattle, goats, sheep, and pigs all have their origins as farmed animals in the Fertile Crescent. Dates for the domestication of these animals range from between 13,000 to 10,000 years ago.[lxx] Agriculture enabled the development of villages, cities, and specialization of labor. Specialization of labor enabled acceleration of improvements in technology. Cities led to the spread of disease more easily and to greater disparity in distribution of property.

Important developments in human history include the development of mathematics (5000 BCE), astronomy and calendars (4000 BCE), the wheel (3500 BCE), and written language. Written language was discovered and used as early as 3400 BCE in Mesopotamia, 3250 BCE in Egypt, and approximately 1200 BCE in China[lxxi]. Harappan script, from the Indus Valley, on objects such as pottery tools, tablets, and ornaments have been dated to around 2800-2600 BCE[lxxii]. Note: Oracle bone script was found engraved in animal bones or turtle plastrons (nearly flat part of the shell structure) dating to the late second millennium BC, and is the earliest known form of Chinese writing.

Oracle Bone Script[lxxiii]

Early writing was used to keep tract of assets and transactions. It progressed to be used for legal records, legal codes, and religious texts. Although the use of money likely began much earlier, the first coins emerged nearly 5000 years ago. The Mesopotamian shekel is the first form of currency known[lxxiv].

Many civilizations have contributed to advancement of technology, culture, and society. Some of the more influential civilizations include[lxxv]:

- Ancient Egypt (3150 – 30 BCE)
- Mayan Empire (2000 BCE – 1540 AD)
- Phoenicia (1200 – 539 BCE)
- Greek Empire (800 BCE – 600 AD)
- Persian Empire (550 BCE – 651 AD)
- Chola Dynasty (350 BCE -1279 AD)
- Chinese Empire (221 BCE – 1912 AD)
- Roman Empire (27 BCE -1453 AD)
- Byzantine Empire (330 – 1453 AD)
- Islamic Golden Age (750 – 1257 AD)
- Mongol Empire (1206 – 1368 AD)
- Ottoman Empire (1299 – 1923 AD)
- Aztec Empire (1428 – 1521 AD)
- Mughal Empire (1526 – 1857 AD)
- British Empire (1583 AD – present)
- United States (1776 AD – present)

As can be seen, most of these civilizations have periods development, growth, and decline. This trend is likely to continue with changes in the most influential civilizations. There are complex and varied reasons for the rise and decline of civilizations.

Presently, there are at least five major civilizations with distinct cultures: Western (including partly Australia and New Zealand), Islamic, Sub-Saharan African, Indian Subcontinent, and Far Eastern. Although distinct and separately evolving major cultures are spread throughout the world, such as South East Asia, South America, Central Asia (including Persia), Japan, Russia, and Polynesia, with mixtures of their own local and some of the five major civilizational cultures.

It is important to note that local cultures throughout history have risen and then collapsed due to a variety of causes including ineffective governments, poor water management, civil wars, foreign invasions, depletion of resources, constant wars of conquest, earthquakes, fires, plagues, overpopulation, volcanic disruptions, and internal rebellions. The current cultures may not persist: some may collapse and vanish, some may evolve or split into new ones and others may mix to form new ones. The current changes to the global climate will unequally affect various civilization cultures, perhaps resulting in the extinction of some of them, as in the past.

The number of technological advances has accelerated during the following years, as evidenced by the scientific revolution (15th – 17th century), the industrial revolution (17th- 18th century), invention of the lightbulb (~1879)[lxxvi], the revolution in electronic communication and information (20th century) and the digital revolution (1980-present),[lxxvii] continuing with artificial intelligence and robotics in the 21st century. The speed of information transmission and amounts information storage have accelerated due to inventions such as the printing press, radio, and computer.

Much of human history is involved in proliferation (spreading of humans across the Earth), obtaining and using resources (i.e., farming, ranching, mining, forestry, industry), government (i.e., military, legal, control), warfare, invention, and construction. Many of these advances have been enabled using increasingly powerful energy sources. The first energy sources used by humans were biofuels such as wood (200,000 BCE)[lxxviii]. Later sources of energy include coal, windmill, oil, solar, hydroelectric, nuclear, and the wind turbine.

Technological advances have led to the proliferation of humans across the Earth and have enabled humans to increase food production, improve travel, increase their life span, and improve the health of humans. Technology has also enabled humans to have significant effects on land, water, air, and other life forms. Technological advances have also been used to advance capabilities for warfare. Potential future advances in technology will be examined in the next chapters.

An overall trend seen regarding the origin of life and evolution is that of expansion and adaptation. Life may have originated around deep-ocean hydrothermal vents, or it may have originated on the Earth's surface at volcanic hot springs. Regardless, life forms have evolved to be able to live in other environments and have expanded from their origin. The Earth has been inhabited by over five billion different species of organism. Most have become extinct, but others now exist. Life forms moved from aqueous environments to land. Life now exists in many extreme environments on Earth, from arid deserts, and frozen tundra, to thermal toxic vents in the deepest reaches of the ocean floor. This trend leads to the expectation that Earthlings will tend to expand to other environments that may include other planets and/or space, depending on the ability to overcome obstacles that would prevent such a spread. We will investigate this further in future chapters.

Evolution of Other Planets

We have discussed the origin of our Universe and have focused on the changes that have occurred on Earth. It is valuable to recognize that stars and other planets have also changed over time. While changes were taking place in our Solar System and on our planet, changes were also taking place across the Universe.

Other planets do not have static conditions, but instead have gone through evolution as has our planet. While conditions on Earth have changed and will continue to change, so is the case with other planets and other components of our Universe. Some planets are much older than the Earth and have gone through significant changes during their existence. Known examples of older planets in the Milky Way Galaxy include[lxxix]:

- PSR B1260-26 b (13 billion years old) in the constellation Scorpius – gas giant
- Keppler-444 planets (11.2 billion years old) in the constellation Lyra – five terrestrial exoplanets
- 55 Cancri e (10.2 billion years old) in the constellation Cancer – believed to be a carbonaceous planet

- HD 80606 b (7.6 billion years old) in the constellation Ursa Major – gas giant
- 51 Pegasi b (6.1-8.1 billion years old) in the constellation Pegasus – gas giant
- Keppler-452b (6 billion years old) in the constellation Cygnus – Super-Earth exoplanet
- TRAPPIST-1 planets (5.4-9.8 billion years old) in the constellation Aquarius – seven terrestrial planets
- Jupiter (4.6 billion years old) in our Solar System – gas giant

Since our spacefaring civilization has taken less than 5 billion years to develop, potentially other alien species could have evolved in our galaxy billions of years before our solar system formed. Since it is in principle possible for aliens to have traveled through the galaxy in one billion years, it would seem possible that we would be visited or colonized by aliens, but it is a paradox that currently, earthlings seem to be alone. This is the so-called Fermi paradox[lxxx]. A lack of aliens is concerning, because among explanations, it may be that intelligent life is rare or that it is extinguished before colonization. Thus, we should be apprehensive about our future survival. An alternative explanation may be that intelligent life is too far away for meaningful two-way communication.

Distant galaxies are likely quite different from what we currently see due to the delay in the time it takes for electromagnetic waves such as light to reach earth, which can take billions of years[lxxxi].

As an example of another planet's evolution, we will examine changes over time in Mars to make this point.

Mars

Mars is an important planet to examine as it is one of our closest neighbors and it may have the greatest opportunity for human inhabitation in the future. Mars has experienced changes in surface conditions that are driven by its thermal evolution, orbital evolution, and by changes in solar input and greenhouse gases[lxxxii]. Mars has a radius that roughly 0.53 of the Earths and it has a smaller mass (approximately 15% of the Earth's mass[lxxxiii]). It has two small moons (Phobos and Deimos with radii of 11.2 and 6.3 km versus our moon, Luna with a radius of 1738 km). These small moons do not protect Mars from space debris and meteors as well as the Earth is protected by our moon.

Mars was unable to retain much air and water over billions of years in part because it lacks a strong magnetic field to protect it from solar wind and thermal evaporation. Today, Mars has weak magnetic fields in various regions of the planet that appear to be the remnant of a magnetosphere. Mars' magnetic fields are approximately16-40 times less than Earth's, which are unable to deflect most of the solar wind's charged particles that strip away the ozone layer and expose it to harmful radiation.

Spacecraft exploration of Mars began in 1965 with a flyby by the Mariner 4[lxxxiv], followed by orbiters, landers, and rovers. This investigation has provided global measurements of topography, geologic structure and processes, surface mineralogy, the near-surface distribution of water, the intrinsic and remnant magnetic field, gravity field, and the atmospheric composition. Surface missions have acquired information on surface morphology, stratigraphy, mineralogy, composition, and atmosphere-surface dynamics. This information has confirmed that Mars has a long and varied history during in which water has played a major role. There remain branching dry channels, showing that rivers once flowed on its surface.

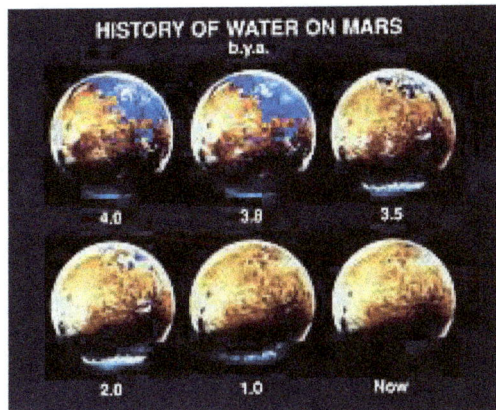

History of Water on Mars (Billion Years Ago)[lxxxv]

Recent exploration has confirmed that the surface of Mars today is cold, dry, chemically oxidizing, and exposed to intense solar ultraviolet radiation. Liquid water might occur episodically near the surface as dense brines in association with melting ice.

The subsurface of Mars has a mean annual surface temperature close to -70 F (-57 C) at the equator and -170 F (-112 C) at the poles. Hydrothermal activity is likely in past or present volcanic areas, and the background geothermal heat flux could drive water to the surface. At depths below a few kilometers, warmer temperatures would sustain liquid water in pore spaces, and a deep-subsurface biosphere is possible, if nutrients are accessible, and water can circulate.

Climate changes in the recent geologic past might have allowed habitable conditions to arise episodically in near-surface environments. Mars undergoes large changes in its obliquity (i.e., the tilt of its polar axis). At present the obliquity ranges from 23° to 27°, with values as high as 46° during the past 10 million years. At these higher obliquities, ground ice is stable closer to the equator and surface ice may be transferred from the poles to lower latitudes.

Recent observations confirm that conditions in the distant past were probably different from present conditions, with wetter and warmer conditions prior to about 3.5 billion years ago. Early Mars

59

had extensive volcanism and high impact rates. The formation of large impact basins likely developed hydrothermal systems and hot springs. A huge crustal crack, the Valles Marineris, indicates that Mars has lost its internal heat convection.

Valles Marineris (more than 4,000 km (2,500 mi) long, 200 km (120 mi) wide and up to 7 km (23,000 ft) deep[lxxxvi]

Since approximately 3.5 billion years ago, rates of weathering and erosion appear to have been low. Analyses of the geomorphology and surface composition of ancient terrains have confirmed that the early Mars climate system was different from today's and that the global environment varied throughout this early period. The magnetic field is weak because of the smaller size of Mars meant that any internal dynamo lost its heat drive early in the planet's history.

A substantial fraction of the exposed terrain of Mars, unlike that of Earth, is inferred to be older than about 3.5 billion to 3.7 billion years. These terrains are represented by topographically high, extensively cratered surfaces that dominate the southern latitudes and provide a geologic record of the early stages of planet formation. Ancient Mars was marked by the presence of near-surface liquid water, with evidence for standing water, lakes, valley networks, and thick, layered sequences of sedimentary rocks with internal stratification. A huge ocean once covered much of the southern hemisphere, which has evaporated in the thin air of the planet to escape into space.

In the last 3.0 billion years Mars appears to have been less active than the planet was in its earlier history, with substantially reduced global aqueous modification and lowered erosion rates. Secondary minerals in Martian meteorites show a range of ages from 3.9 billion to 100 million years old, suggesting that fluids were present in the Martian crust for most of Mars's history.

Since there is water or ice present on Mars, if an atmosphere were present that presented a greenhouse effect, it might be possible to increase temperatures to melt the ice. This presents the concept of terraforming Mars. Terraforming is a hypothetical procedure that would transform Mars from a hostile environment for life to an environment that could sustainably host humans and other lifeforms and could serve as a possible escape from extinction on Earth. The possibility of terraforming Mars will be discussed later (see section on Mars).

Chapter 2 - Current Status and Trends

We will assess the status of the Earth and Earthlings and look at current trends that may influence the future. It is important to determine the causation of the trends to determine whether they are sustainable or not.

Geological Activity

The Earth's outermost layer (lithosphere) is made up of the crust and upper mantle and is broken into tectonic plates. These plates lie on top of a partially molten layer of rock called the asthenosphere. Due to the convection of the asthenosphere and lithosphere, the plates move relative to each other at different rates, from 3-15 cm per year. The continents of the Earth are currently located per the map below on the various tectonic plates described below.

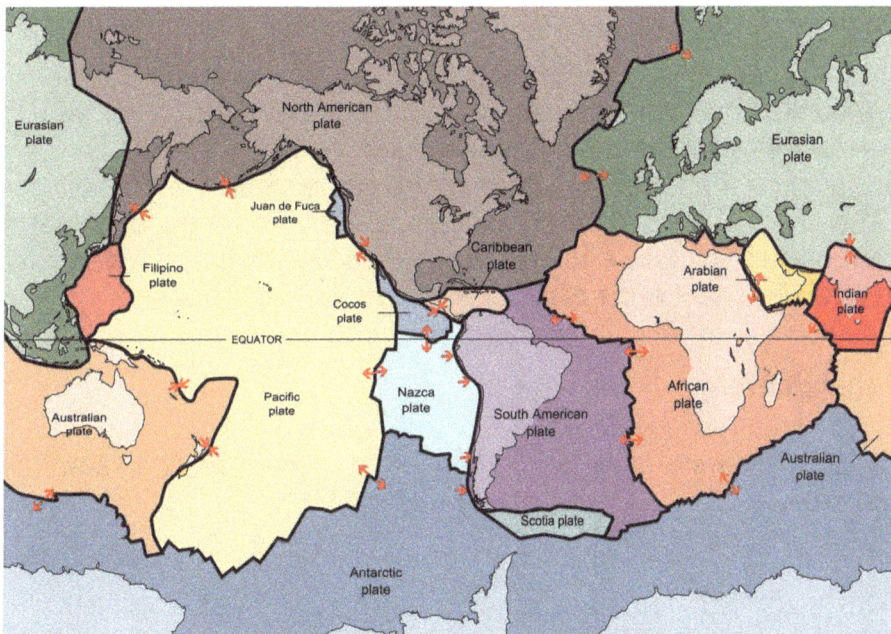

Tectonic Plates in the 20th Century[lxxxvii]

Earthquakes are caused by the Earth's shifting tectonic plates, which float on the planet's molten core in a constant sliding motion. The frequency and magnitude of earthquakes depend upon how the tectonic plates move. Volcanic activity and earthquakes are focused on the plate boundary where two plates come in contact. One such area is an area around the edge of the Pacific Ocean, which generates 75% of the world's volcanos and 80% of the world's earthquakes. We can expect earthquakes and volcanic activity to continue in these areas as the plates continue to move. Supervolcanoes occur when magma in the mantle rises into the crust but is unable to break through it and pressure builds in a large and growing magma pool until the crust is unable to contain the pressure. The impact of volcanic eruptions will depend largely on the extent of the lava flow volume and amount of gaseous and dust emissions. The Lake Toba (North Sumatra, Indonesia - 74,000 years ago) and Flat Landing Brook (New Brunswick, Canada - 466 to 465 Mya) eruptions may have had eruptions covering over 10,000 square kilometers[lxxxviii], resulting in copious injection of gas and dust high into the atmosphere. Since past eruptions have caused mass extinctions and devastation of civilizations, future eruptions will remain a threat to the survival of humans and other lifeforms.

Currently, we can anticipate that certain areas are prone to earthquakes over an expected period. Perhaps in the future we will be able to predict the date, time, location, and magnitude earthquakes. Patterns of seismicity are complex and often difficult to interpret; however, increasing seismic activity is a good indicator of increasing volcanic eruption risk, especially if long-period events become dominant and episodes of harmonic tremor appear.

Global Warming

The Earth's temperature has increased over the past 120 years (see below).

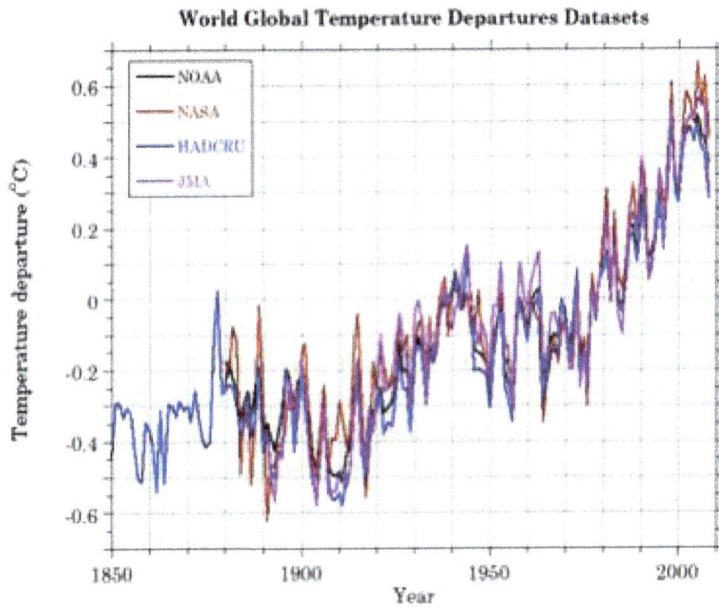

World Global Temperature Departures Datasets

World Global Temperature Departures[lxxxix]

Climate models predict that Earth's global average temperature will continue to rise in the future[xc]. For the next two decades warming of about 0.4° F is projected. If we continue to emit as many, or more, greenhouse gases, this will cause more warming during the 21st Century than we saw in the 20th Century. During the 21st Century, some computer models predict that the earth's average temperature will rise between 3.2° and 7.2° F. The amount of predicted warming differs depending on the quantity of greenhouse gas emissions it assumes for the future. The amount of predicted warming also differs between different climate models. Global warming is predicted to impact regions differently. For example, temperature increases are expected to be greater on land than over oceans and greater at high latitudes than in the tropics and mid-latitudes.

As the climate warms, snow and ice will melt. The amount of summer melting of glaciers, ice sheets, and other snow and ice on land are predicted to be greater than the amount of winter precipitation. The amount of sea ice (frozen seawater) floating in the ocean in the Arctic and Antarctic is expected to decrease over the 21st Century.

65

A warmer climate causes sea level to rise by two mechanisms:
1) melting glaciers and ice sheets (ice on land) add water to the oceans, raising sea level, and
2) ocean water expands as it warms, increasing its volume, and raising sea level.

Global Flooding

During the 20th Century, the sea level rose about 4 to 8 inches (10-20 cm). Thermal expansion and melting ice each contributed about half of the rise, though there is some uncertainty in the exact magnitude of the contribution from each source. By the year 2100, models predict sea level will rise between about 8 to 20 inches (20-50 cm) above late 20th Century levels. Thermal expansion of sea water is predicted to account for about 75% of future sea level rise according to most models. If the sea levels continue to increase many coastal cities will eventually be flooded. Approximately 40% of Americans live along coasts[xci] that may be flooded by the end of the 21st century, causing loss of homes, industry, and farming. Some 40 percent of the world's population lives within 62 miles (100 kilometers) of the ocean, putting millions of lives and billions of dollars' worth of property and infrastructure at risk[xcii]. Sea-level rise can mean that saltwater intrudes into groundwater drinking supplies, contaminates irrigation supplies, or overruns agricultural fields. Low-lying, gently sloping coastal areas are particularly vulnerable to contamination of freshwater supplies.

Global Average Absolute Sea Level Change, 1880–2015

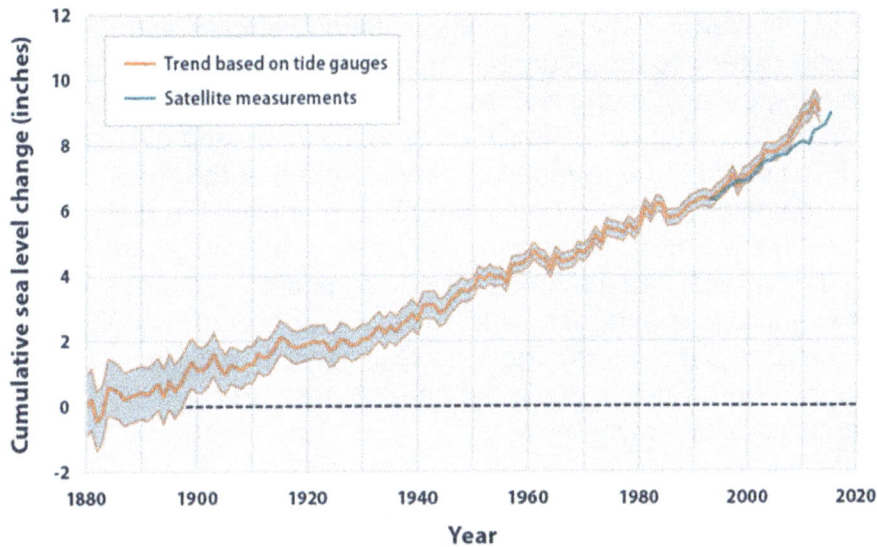

Data sources:
- CSIRO (Commonwealth Scientific and Industrial Research Organisation). 2015 update to data originally published in: Church, J.A., and N.J. White. 2011. Sea-level rise from the late 19th to the early 21st century. Surv. Geophys. 32:585–602. www.cmar.csiro.au/sealevel/sl_data_cmar.html.
- NOAA (National Oceanic and Atmospheric Administration). 2016. Laboratory for Satellite Altimetry: Sea level rise. Accessed June 2016. http://ibis.grdl.noaa.gov/SAT/SeaLevelRise/LSA_SLR_timeseries_global.php.

For more information, visit U.S. EPA's "Climate Change Indicators in the United States" at www.epa.gov/climate-indicators.

Global Average Sea Level Change (1880-2015)[xciii]

Oceanic Disruption

The Earth's oceans are predicted to act as a buffer against global warming by taking up some of the excess heat and carbon dioxide from the atmosphere. This is good news in the short run, but more problematic in the long run. Carbon dioxide combined with seawater forms weak carbonic acid. Scientists believe this process has reduced the pH of the oceans by about -0.1 pH since pre-industrial times. Further acidification changes of -0.14 to -0.35 pH is expected by the year 2100. More acidic ocean water may cause problems for marine organisms. There is evidence that the increase in acidity is impacting the ability of coral reef organisms to form coral, which is leading toward the decimation of coral reefs (seen as calcium carbonate skeletons) and the life forms dependent on them[xciv].

Large-scale ocean currents called thermohaline circulation, driven by differences in salinity and temperature, may also be disrupted as climate warms. Changes in precipitation patterns and the influx of fresh water into the oceans from melting ice can alter salinity. Changing salinity, along with rising water temperature, may disrupt the currents. In an extreme case some thermohaline circulation could be disrupted or even shut down in parts of the ocean, which could have large effects on climate. Thermohaline circulation begins in the Earth's polar regions. When ocean water in these areas gets cold, sea ice forms. The surrounding seawater gets saltier, increases in density and sinks. If global warming reduces the temperature contrast, then the Gulf Stream may weaken or stop. Northern Europe could become much colder, if the North Atlantic circulation weakens.

After almost three more centuries of global warming, the Earth may experience an increase in mean surface air temperature of 17 degrees Fahrenheit. This warming will change wind patterns, melt most of the polar sea ice, and raise the surface temperatures of the world's oceans. These changes will transfer nutrients from the upper ocean to the deep ocean, transforming ocean circulation and phytoplankton growth around Antarctica. Ocean ecosystems may be starved for nutrients and unable to sustain food webs. The phytoplankton that usually form the base of these webs will be less able to engage in photosynthesis[xcv].

Atmospheric Disruption

Warm ocean surface waters provide the energy that drives immense storms. Warmer oceans in the future are expected to cause intensification of such storms. Although there may not be more tropical cyclones worldwide in the future, some scientists believe there will be a higher proportion of the most powerful and destructive storms.

Warmer global temperatures produce faster overall evaporation rates, resulting in more water vapor in the atmosphere and hence more clouds. Different types of clouds at different locations have different effects on climate. Some shade the Earth, cooling climate. Others with their heat-trapping water vapor and droplets.

68

enhance the warming greenhouse effect. Scientists expect a warmer world to be a cloudier one but are not yet certain how the increased cloudiness will feed back into the climate system.

Models of the global carbon cycle suggest that the Earth will be able to absorb less carbon dioxide (CO_2) out of the atmosphere as the climate warms, worsening the global warming.

We have direct measurements of CO_2 concentrations in the atmosphere going back more than 50 years, and indirect measurements (from ice cores) going back hundreds of thousands of years. These measurements confirm that CO_2 concentrations are currently rising well above recent historic levels (see graph below). Human activity is causing the Earth to get warmer, primarily through the burning of fossil fuels, with a smaller contribution from deforestation.

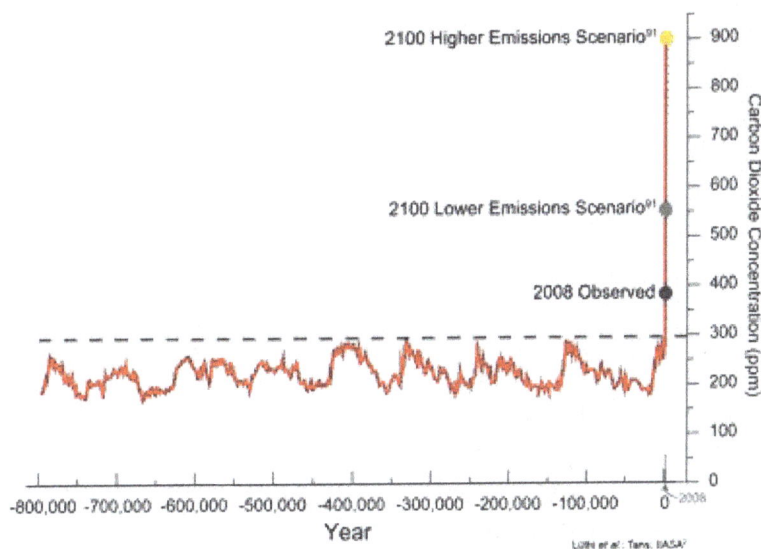

Historic CO_2 Levels[xcvi]

Concentrations of CO_2 in the atmosphere were as high as 4,000 parts per million (ppm) during the Cambrian Period about 500 Mya to concentrations as low as 180 ppm during the Quaternary glaciation of the last two million years[xcvii]. The present concentration is the highest for the last 14 million years.

69

Although the U.S. has one of the highest levels of CO_2 emission per capita (see below), the largest emitter of CO_2 is China (28%), followed by the U.S. (15%), and then India (7%)[xcviii]. Note that the map below is for CO_2 emissions per capita (per person) and while China's and India's total emissions are higher than the U.S., they have higher populations to increase their totals.

The primary source of CO_2 emissions in China is fossil fuels, most notably those that burn coal. The largest sources of CO_2 emissions in the U.S. comes from transportation, industry, and power generation. Even though the U.S. government undertook significant efforts to reduce the reliance on coal for electricity generation, the country has become a major producer of crude oil.

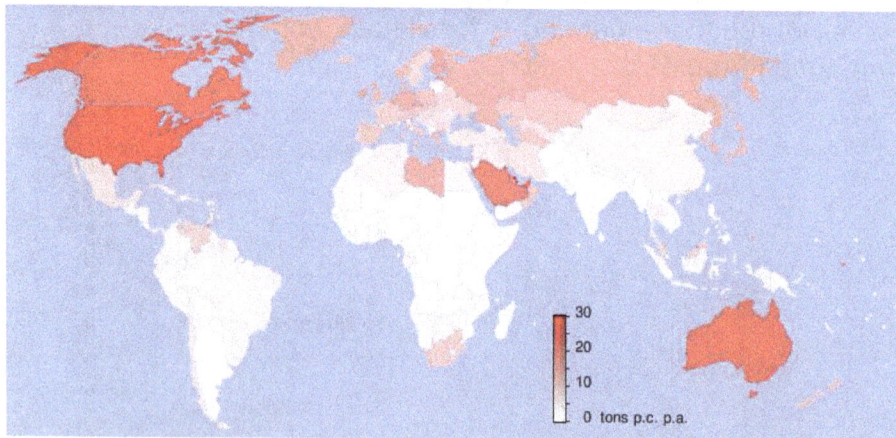

Countries by carbon dioxide emissions per capita[xcix]

Global warming will alter many aspects of biological systems and the global carbon cycle. Temperature changes will alter the natural ranges of many types of plants and animals, both wild and domesticated. There will also be changes to the lengths of growing seasons, geographical ranges of plants, and frost dates, necessitating translocations or migrations for survival.

The greenhouse effect is produced largely by forest clearing and the burning of fossil fuels. It will have the effect of increasing the overall temperature of the Earth.

Earth's Inhabitants or Earthlings

The population of bacteria on the Earth is about 5 million trillion trillion (5×10^{30}). Viruses are estimated to outnumber bacterium by a factor of 10^c. Thus, there are over 10 million times more viruses on Earth than there are stars in the Universe[ci].

Life on Earth in percent by weight[cii].
- Plants: 81.8 %
- Bacteria: 12.7 %
- Fungi: 2.2 %
- Animals: 0.36 %
    - Fish: 0.7 %
    - Humans: 0.01 %
    - Wild mammals: 0.001 %

By weight plants account for most of the life on our planet (approximately 82%). Bacteria significantly outnumber plants, but they weigh less. The current human population is approximately 7.9 billion and humans make up around 0.01% of life on Earth by weight. There are a similar but larger number of bacteria in our bodies as there is of human cells[ciii].

Scientists have identified about 1.9 million species alive today[civ]. They are divided into the six kingdoms of life (see table below[4]).

Scientists are still discovering new species, though the concept of species becomes less applicable to bacteria because many can interchange genes with each other through a process called horizontal gene transfer.

---

[4] In the table below the two kingdoms of Eubacteria (single -celled organism that make up most of the bacteria in the world) and Archaebacteria (simplest of known living cells, often found in harsh environments such as hot springs) are combined.

Note: A species is a group of living organisms consisting of similar individuals capable of exchanging genes or interbreeding.

It is not known for sure how many species really exist today. Most estimates range from 5 to 30 million species. There are approximately 270,000 accepted species of plants. There are estimated to be approximately 1.5 million species of animals, of which around 1 million are insects. Many of the groups of animals that evolved first have a greater diversity, such as insects, mollusks, and crustaceans as compared to birds and mammals. Even fishes, through more evolutionary time, have more species than warm blooded newcomers.

Kingdoms	Described Species	Estimated Total Species
Bacteria (Eubacteria and Archaebacteria)	4,000	1,000,000
Protoctista (algae, protozoa, etc.)	80,000	600,000
Animals	1,320,000	10,600,000
Fungi	70,000	1,500,000
Plants	270,000	300,000
TOTAL	1,744,000	14,000,000

Estimated Numbers of Described Species, and Possible Global Total[cv]

It is noteworthy that the diversity of species largely increases as a function of the distance from the coldest regions to the equatorial regions. Thus, locations such as rain forests and coral reefs (located in warm regions) have greater diversity than regions in the polar areas due to the greater solar energy available for biological processes. Destruction of rain forests and coral reefs will lead to extinction of more species than of most other regions of the Earth. The three most important regions for global biodiversity are South America, the Asia-Pacific, and Central Africa. These tropical megadiverse regions contain approximately 80% of terrestrial biodiversity with a high proportion of threatened species.

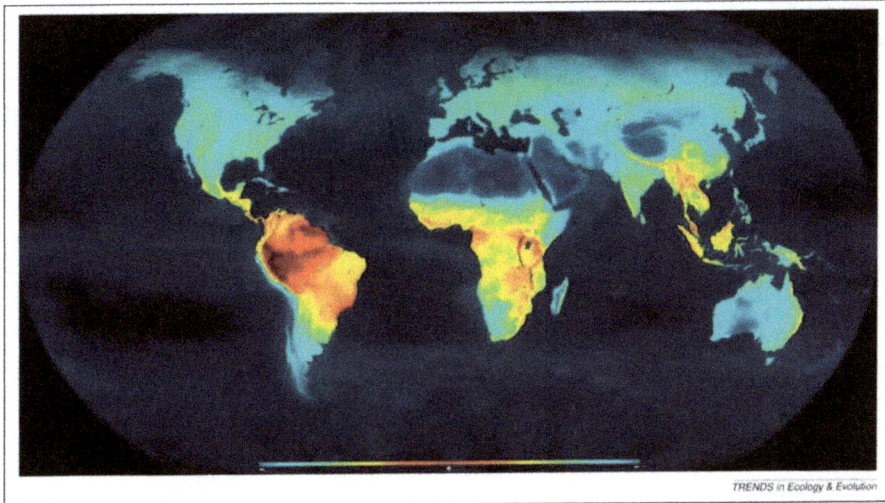

The tropics are 'hotspots' for land biodiversity, as shown in red and yellow. Mannion et al. in Trends in Ecology & Evolution[cvi]

Differences in species are often associated with isolation and differences in climate. Thus, locations such as New Guinea and Australia have been the home to species not found on other continents. Transport of alien species from disparate land masses through human intervention or otherwise has led to subsequent decreases in species variation due to extinction by displacement or competition.

Human Population Growth

Global human population growth is currently around 83 million annually, or 1.1% per year. The global population has grown from 1 billion in 1800 to 7.8 billion in 2021[cvii]. It is expected to keep growing, and estimates have put the total population at 8.6 billion by around 2030, 9.8 billion by around 2050 and between 10.9 to 11.2 billion by 2100[cviii]. Such population growth has been made possible in part due to increasing use of fossil fuels to grow more food and more efficiently deliver it to the population.

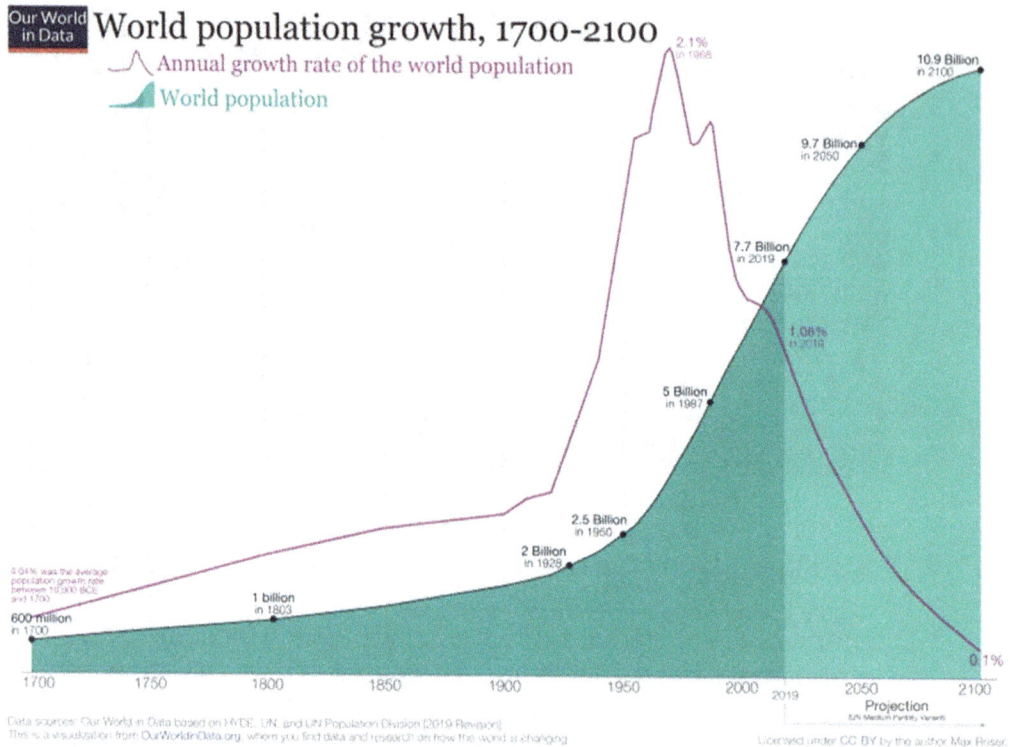

World population growth, 1700-2100

Our World in Data

／‾ Annual growth rate of the world population

▬ World population

2.1%
in 1968

10.9 Billion
in 2100

9.7 Billion
in 2050

7.7 Billion
in 2019

1.08%
in 2019

5 Billion
in 1987

2.5 Billion
in 1950

2 Billion
in 1928

1 billion
in 1803

0.04% was the average
population growth rate
between 10,000 BCE
and 1700

600 million
in 1700

0.1%

1700   1750   1800   1850   1900   1950   2000 2019 2050   2100

Projection
UN Medium Fertility Variant

Past and Predicted World Population Growth[cix]

Long-term estimates of global population suggest a peak at around 2100 of around ten billion people, and then a slow decrease, but there is a lot of uncertainty regarding these numbers. Even if the there is an eventual plateau or decrease in human population, it is unclear whether such a large population will be sustainable (this will be discussed later – see sections on Sustainability and Human Population Growth).

Drivers of population growth include fertility rate, mortality (death) rate, and migration. Fertility rate is a measure of the number of children on average that a woman will bear in her lifetime. The global average fertility rate is just below 2.5 children per woman today. In the pre-modern era fertility rates of 4.5 to 7 children per woman were common in part due to the high death rate in infants and young children. Reasons for this change are likely related to the empowerment of women (increasing access to education and increasing labor market participation), declining child mortality due

74

to modern medicine, modern birth control methods, and a rising cost of bringing up children (to which the decline of child labor contributed). Higher child survival rates and increased urbanization all the lower need for children to contribute to family farming. Time intensive child raising reduces the number of children that parents may want. Methods of contraception give parents the chance to limit fertility closer to their desired fertility. The trend continues despite new procedures to lead to fertility (intravaginal insemination, in-vitro fertilization, etc.). Other factors may involve religious beliefs, government intervention, and cultural norms.

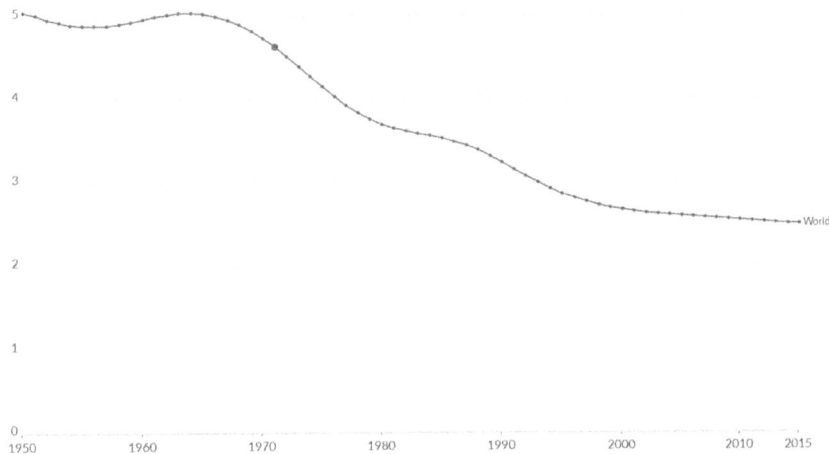

Children per Woman 1950-2015[cx]

There has also been a noticeable drop in sperm counts of males and in the fertility rates of females recently[cxi]. Sperm counts have decreased by 50% in the past forty years. The decline is occurring in Western countries, and it is continuing without signs of tapering off. This trend may be in part due to increased chemicals in our food and environment, although modern lifestyles (diet, stationary jobs, stress) may also be contributing. Some of the chemicals involved are antiandrogens and chemicals with estrogenic effects. Endocrine disrupting chemicals are included in plastics and cosmetics (such as phthalates, bisphenol A), flame retardants, and pesticides. Exposure to these chemicals can be disruptive to the reproductive system during childhood, adolescence, and adulthood, detrimental to the fetus during development, and can be transferred through milk during breastfeeding. These environmental toxic effects are not only affecting human fertility but are also impacting the fertility and well-being of other animals and life-forms.

It is worth noting that the fertility rate varies over countries. Many countries in Africa have higher fertility rates than those of other continents.

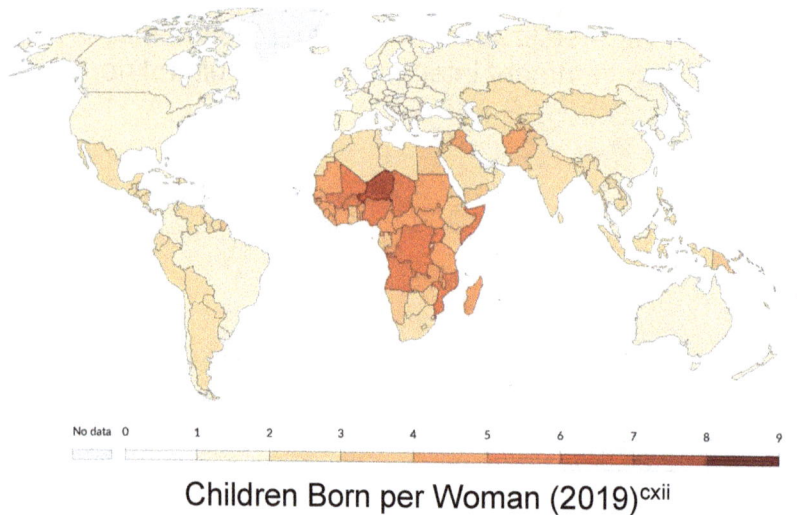

Children Born per Woman (2019)[cxii]

Mortality rate or death rate is a function of deaths per population and life expectancy. Current leading causes of death include ischemic heart disease, stroke, chronic obstructive pulmonary disease, lower respiratory diseases, lower respiratory infections, cancer, and neonatal conditions[cxiii]. Mortality rates can be obtained by vital statistics and census data. World crude death rates (death from all causes) per 1000 people have decreased from around 19 in 1950 to around 8 in 2020. Decreased mortality rates in recent history are driven by factors such as: modern medicine, better housing, and smaller family sizes relating to improved living standards. The crude death rate is expected to increase to around 10 in 2050 as the world becomes more crowded.

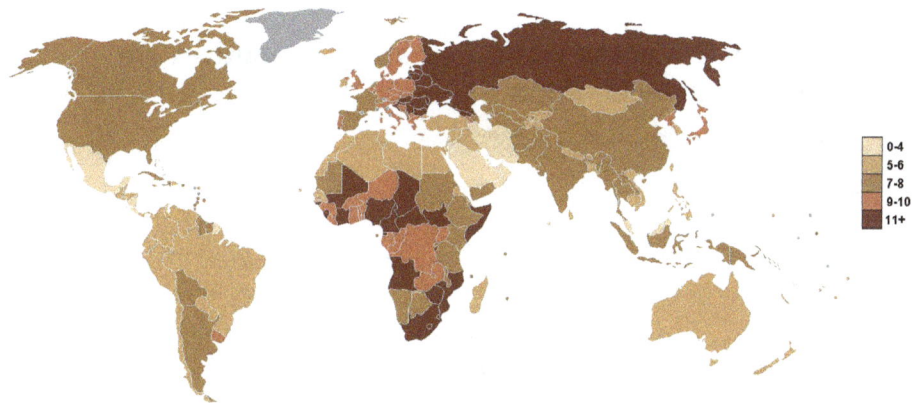

Death Rates of Countries (2000-2005) –
Deaths per thousand per year[cxiv]

While China and India will continue to dominate the total
population in terms of sheer numbers, African nations will see the
sharpest increases in population, accounting for more than half of
the global population growth over the next 35 years. All of the 48
least-developed countries, of which 27 are in Africa, are expected
to witness steep population growth. Nigeria is expected to emerge
as the third most populous country by 2100.

By 2100, five of the world's 10 largest countries are projected to be in Africa

Countries with largest population, in millions

1950		2020		2100	
China	554	China	1,439	India	1,450
India	376	India	1,380	China	1,065
U.S.	159	U.S.	331	Nigeria	733
Russia	103	Indonesia	274	U.S.	434
Japan	83	Pakistan	221	Pakistan	403
Germany	70	Brazil	213	D.R. Congo	362
Indonesia	70	Nigeria	206	Indonesia	321
Brazil	54	Bangladesh	165	Ethiopia	294
UK	51	Russia	146	Tanzania	286
Italy	47	Mexico	129	Egypt	225

Note: Countries are based on current borders. In this data source, China does not include Hong Kong, Macau or Taiwan.
Source: United Nations Department of Economic and Social Affairs, Population Division, "World Population Prospects 2019."

PEW RESEARCH CENTER

Predictions of the 10 Largest Countries[cxv]

Migration can have a significant impact on population change. Africa remains the continent with the highest number of people ready to migrate for economic, conflict, or climate reasons or simply to seek better life perspectives. The rate and migration trends will depend on major events such as wars, poverty, job opportunities, and resources.

Increased human populations are expected to cause increased demand for resources such as fresh water and food, consumption of natural resources (such as fossil fuels) faster than the rate of regeneration, and a deterioration in living conditions. It may be that a peak in human population will be reached and that a subsequent decrease will be seen. Unless there are significant decreases in the population the sustainability of meeting the resource demands of humans is questionable.

Probable Shortages

In understanding the effects of over-consumption of planetary resources, it is important to understand that there are a variety of goods and services that the world population constantly consumes. These range from food and beverage, clothing and footwear, housing, energy, transportation, education, health and personal care, financial services, and utilities such as water and sewage. Some key shortages that are likely to develop in the next century are water, food, and energy shortages. Currently, most food production, manufacturing, and distribution is energy intensive, so when energy shortages occur, food shortages also occur[cxvi].

Food

Currently, 75% of the world's food is generated from only 12 plants and 5 animal species.

75% of the World's Food
Comes from 12 Plants and 5 Animal Species

Rank	Top Plants	Annual Global Production Metric Tons, 2011	Annual Global Value in USD Billions, 2012
1	Sugar	1,800,377,642	$57
2	Maize	885,289,935	$55
3	Rice	740,961,445	$337
4	Wheat	701,395,334	$84
5	Potatoes	373,158,351	$50
6	Soy Beans	262,037,569	$65
7	Cassava	256,404,044	$25
8	Tomatoes	159,347,031	$58
9	Banana	107,142,187	$29
10	Onions	86,343,822	$18
11	Apples	75,484,671	$32
12	Grapes	69,093,293	$39
Rank	Top Animals	Annual Global Production	Annual Global Value
1	Beef & Milk	725,123,869	$622
2	Chicken & Eggs	155,183,059	$182
3	Pork	118,168,709	$306
4	Goat Milk & Meat	20,000,000	n/a
5	Sheep	8,229,068	$22

Source: FAO via Wikipedia (http://bit.ly/2QAkmhC)

75% of the World's Food[cxvii]

Interestingly, insects appear to be a largely unused source of food for human consumption.

Food production has continued to increase to meet the increase in population. Also, population has increased to match food production. There have been regional shortages that have led to starvation, malnutrition, increased mortality, and conflict. These same types of effects are seen when food shortages are seen with other life forms besides humans.

Many animals reproduce in high numbers. Arthropods and fishes can lay from thousands to millions of eggs during their lifetime. If most of these animals survived, then animal populations would grow rapidly and exponentially. This is not what happens. Animal populations tend to stay relatively stable across generations. For a population to remain stable, on average only one offspring per parent can survive to adulthood. The rest will die. There are many causes of death, including death by starvation after being born or hatched. Sometimes the effects of hunger and malnutrition are reduced because malnourished females do not get pregnant.

For thousands of years since the agricultural revolution, the global population was able to feed itself comfortably but there have been many localized famines. Some of the worst famines were the Great Famine in China (1959-1961) due to a forced change in workers from farming to mining. The famine in China in 1907 was largely caused by heavy rains and floods. The Famine of the Skull (1788-1794) in India, the Bengali Famine in 1770, and the famine in Persia (1917-1919), famine in Russia (1921-1922), Great Famine in Europe (1315-1317) were due to factors including colonization, crop failures, war, and overpopulation[cxviii]. In the 17th century Europe, overpopulation compared to food may have contributed to wars, revolutions, and migrations to the Americas. More localized famines can be expected.

In the early nineteenth century, the global population had significantly increased leading to unrest, which was mitigated by the Industrial Revolution with access to more energy provided by coal and oil. At around the same time, we began to burn fossil fuels in large quantities resulting in increasing carbon dioxide in

80

the atmosphere, which has contributed to global warming. Fossil fuels have enabled industrial agriculture. Fossil fuels are used to make fertilizers and pesticides, run harvesting engines, vehicular transportation, and refrigeration of food. Currently, we are living in the most densely populated and heavily polluted societies in history.

As populations continue to grow and our capacity for pollution continues to increase, our food systems are becoming increasingly vulnerable. Agriculture requires fertile soil, water, energy, and clean air to remain healthy and productive. Each of those necessities are jeopardized by overpopulation, depletion of energy reserves, and human-driven global warming.

Wars have caused food shortages by the destruction of farmland and farm equipment, by hampering workers through safety threats and conscription, commandeering transport routes, and destruction of transportation lines and vehicles.

Humans have and will continue to work to address food shortages by improvements in technology. Food shortages may be addressed by genetically engineered food (providing higher crop yields) or other agricultural advances. Robotic farming may be helpful in labor saving and efficiency[cxix]. Ultimately, any increases in human population will need to be balanced by a sustainable food production. Alternatively, if the human population were to decrease, the challenge of a sustainable food supply would be more easily achievable.

Water

Water covers approximately 70% of our planet. However, freshwater, which we drink, bathe in, and irrigate farm fields is about 3% of the world's water. Two-thirds of the fresh water is trapped in frozen glaciers or is otherwise unavailable for our use. As a result, over one billion people worldwide lack adequate access to water, and 2.7 billion find water scarce for at least one month of the year. Inadequate sanitation, which requires water, is also a problem for 2.4 billion people[cxx]. Albeit coastal areas could use seawater for sanitation.

Many of the water systems that keep ecosystems thriving and feed a growing human population have become stressed. Rivers, lakes, and aquifers are drying up or becoming too polluted to use. More than half the world's wetlands have disappeared. Agriculture consumes more water than any other source and wastes much of that through inefficiencies. Global warming from excess atmospheric carbon dioxide is altering patterns of weather and water around the world, causing shortages and droughts in some areas, and floods in others. With the use of additional energy some of these localized disparities could be mitigated through storage and pipeline transport.

At the current water consumption rate, scarcity will worsen. It is estimated that by 2025, 40-66% of the world's population may face water shortages[cxxi]. According to the U.S. Intelligence Community Assessment of Global Water Security, by 2030 humanity's annual global water requirements will exceed current sustainable water supplies by 40%[cxxii].

Between 2050 and 2100, there is an 85 percent chance of a drought in the Central Plains and Southwestern United States lasting 35 years or more.[cxxiii] Currently in 2022, a several year drought has depleted major reservoirs in the American Southwest such as Lake Meade and Lake Powell[cxxiv]. Pressure on water resources is increasing in many parts of the world including China, India, Pakistan, the Middle East, and regions of Africa.

The growing scarcity of water on Earth is a danger to humans and nature. The water shortage will also impact food production and our ability to feed the growing population. Having access to water has become a powerful global economic issue that could become one of the main causes of international tension. Local conflicts, sometimes resulting in warfare, are triggered over scarce water resources. Historic and future areas of water conflict include the Middle East (Euphrates and Tigris River conflict among Turkey, Syria, and Iraq; Jordan River conflict among Israel, Lebanon, Jordan, and the Palestinian territories), Africa (Nile River conflict among Egypt, Ethiopia, and Sudan), Central Asia (Aral Sea conflict among Kazakhstan, Uzbekistan, Turkmenistan, Tajikistan, and Kyrgyzstan), and south Asia (Ganges River conflict between India and Pakistan)[cxxv].

Freshwater stress

Global Freshwater Stress[cxxvi]

Efforts to prevent water scarcity include improving water infrastructure, reclaiming water (rainwater harvesting and recycling wastewater), efficient irrigation, conservation, desalination, reduction of pollution, and better sewage treatment. Currently desalination is not cost effective in many locations due to the high energy cost to provide heat for this process. However, in desert areas such as in the Middle East (such as Israel, Saudi Arabia) with much sunlight, located near to seas, desalination has proven possible.

Fujairah Desalination Site in the United Arab Emirates[cxxvii]

Energy

Humans use more than a million terajoules of energy daily.[cxxviii] World electricity production in 2015 was obtained largely from fossil fuel (66%), nuclear power (11%), and hydroelectric (16%).

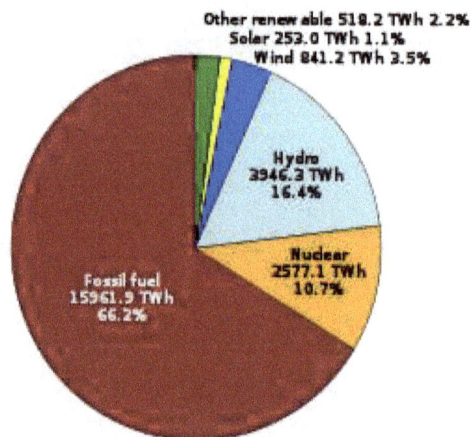

Other renewable 518.2 TWh 2.2%
Solar 253.0 TWh 1.1%
Wind 841.2 TWh 3.5%

Hydro
3946.3 TWh
16.4%

Nuclear
2577.1 TWh
10.7%

Fossil fuel
15961.9 TWh
66.2%

World Electricity Generation by Source (2015)
World Electricity Generation[cxxix]

With the global population increasing and industrialization on the rise in developing nations, energy requirements are anticipated to continue to rise. Since commercial oil drilling began in the 1850s, we have used more than 135 billion tons of crude oil to drive our cars, fuel our power stations, and heat our homes.

Burning of coal, oil, and gas has been linked to the rising levels of greenhouse gases in the Earth's atmosphere and is a leading contributor of global warming. It is anticipated that we are on a path towards energy shortages and environmental changes that can only be stopped by significantly decreasing the use of fossil fuels.

Efforts are being made to move away from the use of fossil fuel, which will increasingly become necessary. Unfortunately, some of the current technologies, such as solar and wind energy generation, provide energy intermittently and current storage technologies are costly and inefficient. Many storage technologies do exist, such as gravity, pressure, or battery storage, so dwindling energy supplies and high demand will make storage worthwhile to implement. Also, the generation of solar panels, batteries, and wind turbines requires mining and energy for manufacturing.

Traditional fission nuclear power is being used to provide power and is more reliable but has pollution challenges with storage of its waste products, which has some fission products with half-lives of over 100 thousand years. Potential fusion power and less polluting thorium fission power provide hope that eventually new abundant energy supplies will become available.

Where at lower cost, geothermal energy has and is being developed on hot spots in the Earth's crust, where magma lies close to the surface. While it is costly to drill into these areas, it has the potential to provide more reliable long-term energy to supplement other energy sources.

The use of electric cars and trucks has potential for decreasing the need for gasoline and diesel fuel, but the vehicles will require electricity, which can be obtained from the electrical grid. If the electrical grid becomes powered using renewable energy (instead of natural gas, coal, and oil), this could be helpful.

Ultimately, if power requirements do not decrease (they are expected to increase), we will need to find alternative sources of energy and improve the efficiency of these new technologies. Technologies such as obtaining energy from fusion may provide a potential solution. This will be discussed more later (see section on Energy).

Sustainability

Sustainability is the ability to avoid depletion of natural resources to maintain an ecological balance. Ultimately, sustainability is not possible (see example of the Sun's evolution due to fusion) but processes of recycling can enable reuse of resources. Challenges to sustainability include environmental degradation, overconsumption, population growth, and economic growth.

Extension of the ability to maintain sufficient water, food, energy, etc. can be achieved by reducing the demand (potentially by reducing the population), recycling, and finding alternative sources, which may have to come ultimately from outside our planet. We have a net input of energy from the Sun, which can be used to provide energy for work (i.e., efficient solar with minimal environment impact). Conversion of mass to energy has potential for improving sustainability, if it can be done in such a way that there is a net energy gain with minimal environmental impact (i.e., fusion).

A driver of human impact on Earth systems is the destruction the Earth's ecosystems. The environmental impact of a community or of humankind depends both on population and impact per person, which depends on what resources are being used, whether those resources are renewable, and the scale of the human activity relative to the carrying capacity of the relevant ecosystems.

Pollution

Humans are now responsible for major new changes in the constituents of the atmosphere, water, and land. When these changes are considered detrimental, we consider the outcome to be pollution. Pollution can take the form of chemical substances, heat, or light. Pollutants, the components of pollution, can be either foreign substances, energies, or naturally occurring contaminants. The recent increased population and industrial expansion have led to an increase in the pollution of our planet, although there have been some recent efforts to clean up some of our environment.

Pollution is a problem mainly because it disturbs biochemical or reproductive processes that are fundamental to many life functions. Major forms of pollution include: air pollution, light pollution, littering, plastic pollution, soil contamination, radioactive contamination, thermal pollution, and water pollution. Some of these pollutions will be discussed in the following sections on the atmosphere, water usage, and land usage. A new source of pollution in Space is high-speed fragments of satellites and rockets that pose collision hazards.

The Atmosphere

Our atmosphere is a layer of gases that surround the Earth extending to about 10,000 km. The layers of the atmosphere are described below.

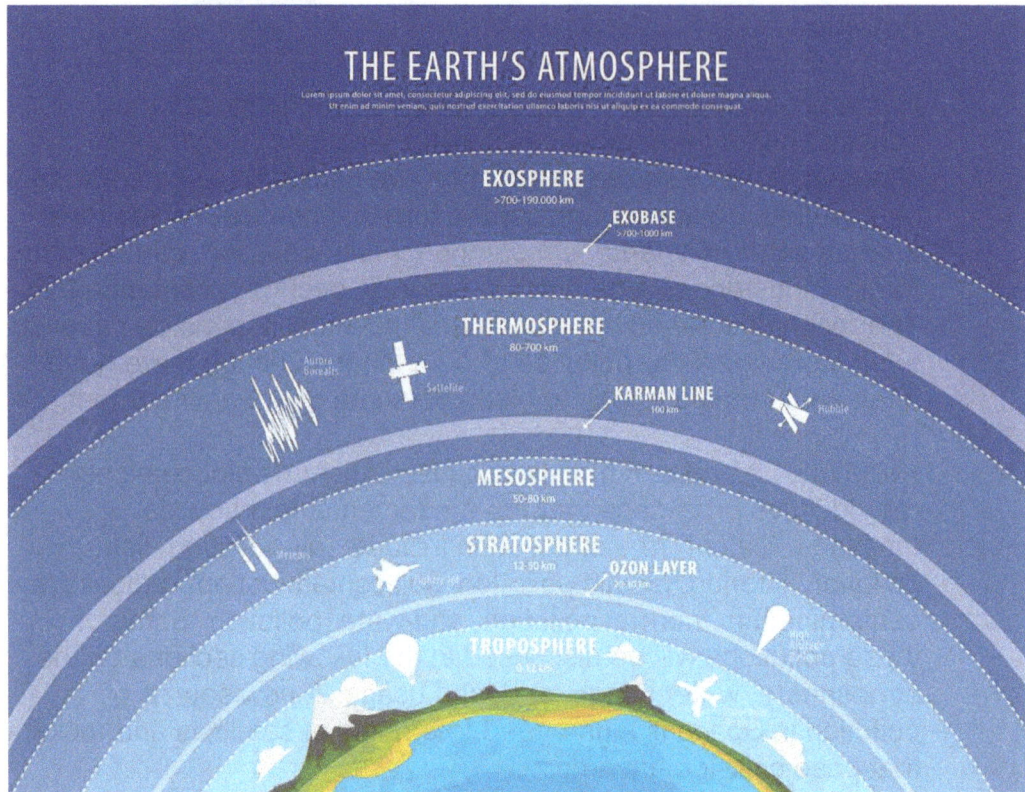

The Earth's Atmosphere[cxxx]

Troposphere

The troposphere starts at the Earth's surface and extends about 8 to 14.5 kilometers high (5 to 9 miles). This part of the atmosphere is the densest. Almost all weather is in this region. Tropospheric air pollution includes ozone precursors, airborne toxics, polycyclic aromatic hydrocarbons, and particles. Tropospheric air pollution has impacts on scales ranging from local to global.

Stratosphere

The stratosphere starts just above the troposphere and extends to 50 kilometers (31 miles) high. The ozone layer, which absorbs and scatters the solar ultraviolet radiation, is in this layer. Pollutants can reach the stratosphere in two ways:
  (1) by direct injection, as from volcanic eruptions, supersonic transports, military aircraft, rockets, or nuclear bombs, and

(2) by indirect injection from the troposphere, which may occur for inert, water-insoluble substances that will eventually work their way up into the stratosphere.

Volcanic emissions that reach the stratosphere can block sunlight, cooling and shading the Earth to cause crop failures in the case of extreme eruptions.

Mesosphere

The mesosphere starts just above the stratosphere and extends to 85 kilometers (53 miles) high. Meteors typically burn up in this layer.

Thermosphere

The thermosphere starts just above the mesosphere and extends to 600 kilometers (372 miles) high. Auroras occur and some satellites orbit in this layer. The average density of air in this region has dropped recently dropped by around 10%, which may be attributed to effects of increased CO_2.[cxxxi]

There are currently around 22,000 objects in orbit right now. The majority of which are waste materials that "pose serious damage to satellites and spacecrafts." These materials include leftover boosters and parts that get ejected from space shuttles during launches, as well as antiquated, broken-down satellites[cxxxii]. If this continues, a cloud of satellite and rocket space junk may become a problem for space travel. Collisions of fragments can generate smaller debris, which could confine humans inescapably to Earth because any spaceship attempting to travel through this area would be disabled or obliterated by countless speeding bullets of space trash.

Exosphere

This is the upper limit of our atmosphere. It extends from the top of the thermosphere up to 10,000 km (6,200 mi). The lower part of the exosphere is the exobase. The exobase can range anywhere from about 500 to 1000 km into the atmosphere, depending on the intensity of solar and geomagnetic activity.

The constituents of our atmosphere have changed over geologic time. The amount of vegetation has an influence in the concentration of carbon dioxide and oxygen. Current changes in our atmosphere are largely caused by human actions. The burning of fossil fuels for transportation and electricity, largely in the U.S, China, and India (consuming 54% of the world's fossil fuels by weight[cxxxiii]), is one of the biggest sources of air pollution. While there are some efforts to reduce air pollution in some Western countries, the industrialization of China and India is likely to continue increases in air pollution.

The fumes from car exhausts contain gases and particulates including hydrocarbons, nitrogen oxides, and carbon monoxide. These gases rise into the atmosphere and react with other atmospheric gases, creating toxic gases. Some major air pollutants known to harm human health include particulate matter, nitrogen dioxide, sulfur dioxide, ozone, and carbon monoxide. These pollutants are released through human activity, as well as those that occur naturally such as desert dust, sea spray, and volcanic emissions. The quality of the air in the era of fossil fuels is expected to worsen, increasing respiratory diseases.

The use of fertilizers for agriculture is a major contributor of fine-particulate air pollution, with most of Europe, Russia, China, and the United States being affected. The level of pollution caused by agricultural activities is thought to outweigh all other sources of fine-particulate air pollution in these countries. Nitrogen in fertilizers also can wash away into waterways, polluting rivers, lakes, and oceans.

Ammonia is the primary air pollutant that comes from agricultural activities. Ammonia enters the air as a gas from concentrated livestock waste and fields that are over fertilized. This gaseous ammonia combines with other pollutants such as nitrogen oxides and sulfates created by vehicles and industrial processes, to create aerosols. Aerosols are tiny particles that can penetrate the lungs and cause heart and pulmonary disease.

Water Usage

A growing global population and shifts towards more resource-intensive consumption patterns has led to an increase in global freshwater use. Freshwater withdrawals for agriculture, industry, and municipal uses have increased nearly six-fold since 1900. Rates of global freshwater use increased sharply from the 1950s onwards, but since 2000 it appears to be plateauing, or at least slowing. The US uses more than 500 billion liters of freshwater every day to cool electric power plants, and roughly the same amount is needed to irrigate crop fields[cxxxiv]. Thermal water pollution from ejected industrial hot water into ponds, lakes, and rivers kills wildlife and destroys ecosystems.

Water pollution can come from a variety of sources. Pollution can enter water through discharges from factories or imperfect water treatment plants. Spills and leaks from oil pipelines or hydraulic fracturing operations can degrade water supplies. Wind, storms, and littering can also send debris into waterways, particularly any plastics, which are often toxic and break down into particles that are absorbed by animals and then humans.

The main cause of water quality problems is from pollutants carried across or through the ground by rain or melted snow. Such runoff can contain fertilizers, pesticides, and herbicides from farms and homes; oil and toxic chemicals from roads and industry; sediment; bacteria from livestock; pet waste; and other pollutants. When electronic waste is dumped into landfills, toxic materials can seep into the soil and groundwater, affecting not only our health, but also land and sea animals.

Pipes can pollute drinking water, if the water is not properly treated. While steps have been taken to reduce exposure to lead in tap water (including the 1986 and 1996 amendments to the Safe Drinking Water Act), lead in water can come from homes with lead services lines that connect to the home. Another drinking water contaminant, arsenic, can come from naturally occurring deposits and from industrial waste.

Water pollution can result in human health problems, poisoned wildlife, and long-term ecosystem damage. When agricultural and industrial runoff with excess nutrients such as nitrogen and phosphorus feed into waterways, these nutrients often fuel algae blooms. The algae blooms create dead zones, or low-oxygen areas, where fish and other aquatic life can no longer thrive. Some bacterial blooms excrete toxins into the water, contaminating and killing animals, posing a hazard to humans.

Algae blooms can create health and economic effects for humans, causing rashes and other ailments. High levels of nitrates in water from nutrient pollution can also be particularly harmful to infants, interfering with their ability to deliver oxygen to tissues and potentially causing "blue baby syndrome[cxxxv]."

Globally, unsanitary water supplies also exact a health toll in the form of disease. At least 2 billion people drink water from sources contaminated by feces and that water may transmit diseases such as cholera and typhoid.

Regulation of pollutants is subject to changing political policy. In developing countries, around 70% of solid waste is dumped directly into the ocean or sea. This causes problems including the harming and killing of sea creatures, which ultimately affects humans[cxxxvi].

Land Usage

Due to population and economic growth, urban sprawl, and the quick expansion of industrial and agricultural activities, land use has experienced changes worldwide over the past half century. Land across the world is being converted from wilderness to land for use in agriculture.

Land use over the long-term, World, 0 to 2016
Total land area used for cropland, grazing land and built-up areas (villages, cities, towns and human infrastructure).

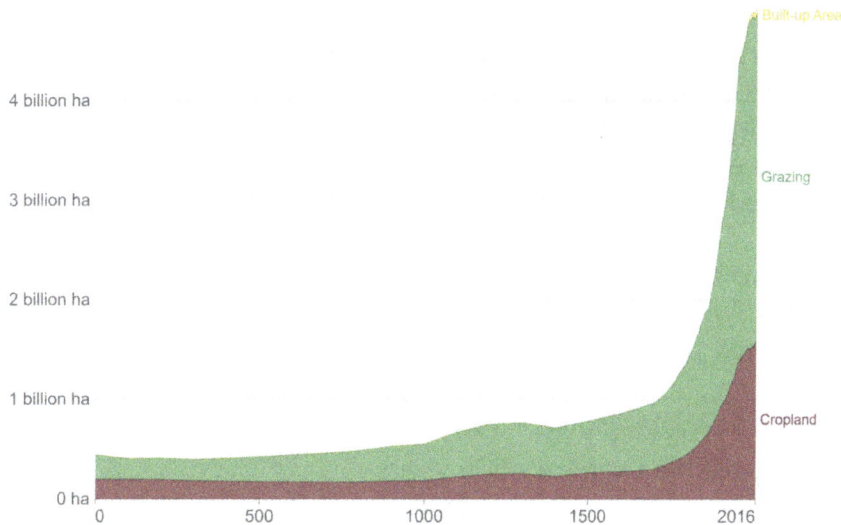

Source: History Database of the Global Environment (HYDE)   OurWorldInData.org/land-cover/ • CC BY

Land Use Over Time[cxxxvii]

Ten percent of the world is covered by glaciers, and 19% is barren land – deserts, dry salt flats, beaches, sand dunes, and exposed rocks[cxxxviii]. This leaves what we call habitable land. Half of all habitable land is used for agriculture. This leaves only 37% for forests; 11% as shrubs and grasslands; 1% as freshwater coverage; and the remaining 1% is built-up urban areas that includes cities, towns, villages, roads, and other human infrastructure.

Chemicals such as pesticides and herbicides are applied to the soil in many farmlands. Industrial and personal waste products are in some cases polluting the land. Soil is polluted through leaking underground septic tanks, sewage systems, the leaching of harmful substances from landfill, and direct discharge of wastewater by industrial plants into rivers and ocean.

Mining activities result in both chemical and toxic heavy metal pollution to soil and water immediately adjacent to an extraction site, and secondary impacts, which occur because of infrastructure development, population movements, and changes in local economies. Secondary impacts can include deforestation along roadways constructed for transporting the extracted products, and biodiversity loss, e.g., through increased fuel wood harvesting, bushmeat hunting, or poaching. They can also include the introduction of invasive non-native species through transport operations, expanded agriculture into natural forests, and expanded logging. Clear-cutting of trees can damage streams and kill off river ecosystems causing the freshwater fishing industry to be damaged.

Landfills are being overloaded with plastic products that can take hundreds of years or more to biodegrade. Below is the lifespan of some popular plastic products[cxxxix]:

- Plastic Water Bottle - 450 years
- Disposable Diapers - 500 years
- Plastic 6-Pack Collar - 450 Years
- Extruded Polystyrene Foam - over 5,000 years

Some bacteria have recently evolved to consume some plastics, potentially shortening the decay times listed above to more sustainable durations, though this effect is still in research.

Monitoring

Modern satellite, rocketry, and sensor technologies have developed, which enables monitoring of the Earth for many research subjects including meteorology, oceanography, terrestrial ecology, glaciology, atmospheric science, hydrology, and geology. Satellites in a variety of orbits of different altitudes and tilts enable collection of information using measurements of emitted energy over various portions of the electromagnetic spectrum (e.g., ultraviolet, visible, infrared, microwave, or radio). This provides information about the Earth's global systems to be observed, monitored, and modelled, allowing better predictions, management of resources, and solutions to local and global problems.

Chapter 3 – The Near Future

Societal Changes

The standard of living over human history has changed from largely poverty, illiteracy, and poor health, to increased average prosperity, literacy, and a longer life. Although there is a large increase in the human population, the number and percentage of people in extreme poverty has decreased (see below). Extreme poverty is defined as living on less than $1.9 per day (adjusted for price differences between countries and for price change over time).

World population living in extreme poverty, World, 1820 to 2015

Extreme poverty is defined as living on less than 1.90 international-$ per day.
International-$ are adjusted for price differences between countries and for price changes over time (inflation).

Our World in Data

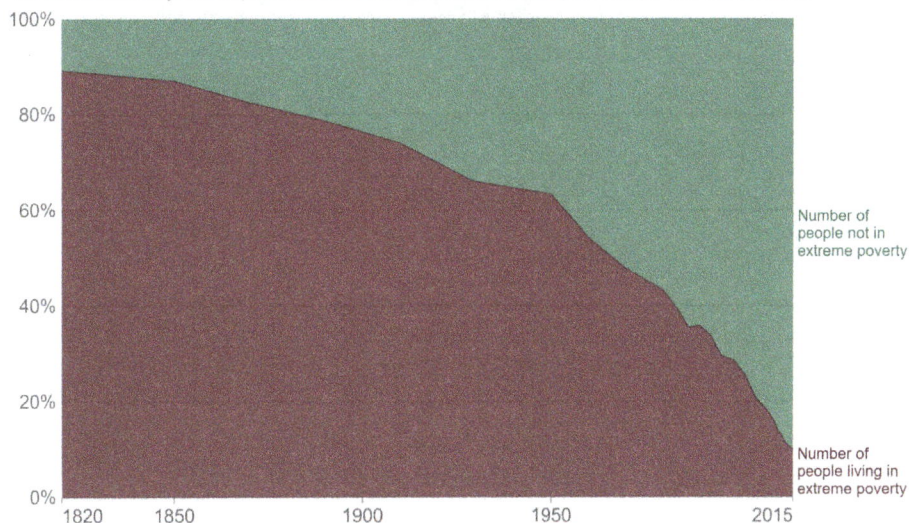

Number of people not in extreme poverty

Number of people living in extreme poverty

Source: Ravallion (2016) updated with World Bank (2019)
OurWorldInData.org/extreme-poverty/ • CC BY
Note: See OurWorldInData.org/extreme-history-methods for the strengths and limitations of this data and how historians arrive at these estimates.

World Population Living in Extreme Poverty[cxl]

The number of and percentage of literate people (above the age of 15) has increased over the past 200 years. From a historical perspective, the world went through a great expansion in education over the past two centuries. This can be seen across all quantity measures. Global literacy rates have been climbing over the course of the last two centuries, mainly though increasing rates of enrollment in primary education. Secondary and tertiary education have also seen drastic growth, with global average years of schooling being much higher now than a hundred years ago. Despite all these worldwide improvements, some countries have been lagging, mainly in sub-Saharan Africa, where there are still countries that have literacy rates below 50% among the youth[cxli].

Literate and illiterate world population

Population 15 years and older.

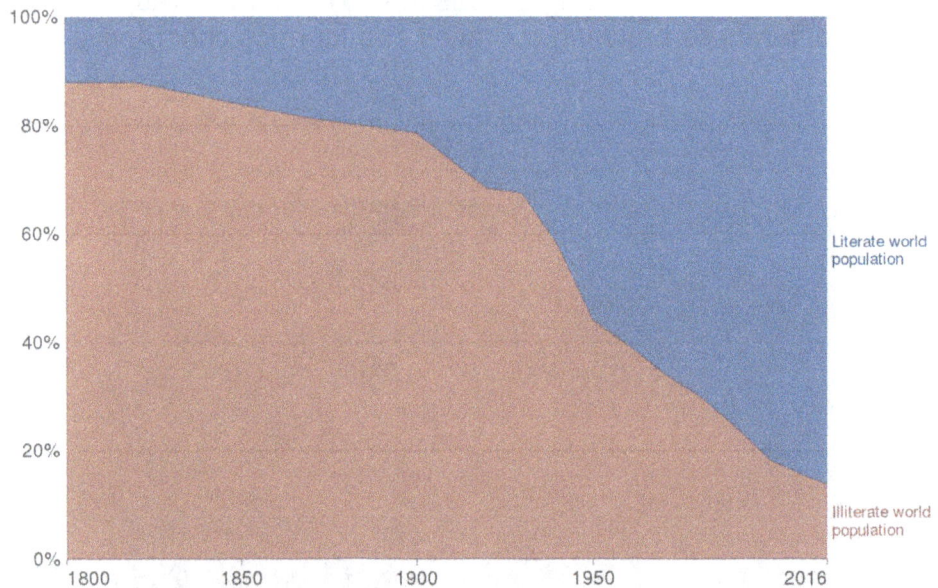

Literate and Illiterate World Population[cxlii]

Shrinking poverty and growing literacy are interrelated by complex feedback in which the knowledge contained in writing allows for humans to become more productive, efficient, and wealthier, permitting more free time to read and write, increasing literacy even more in a positive feedback loop.

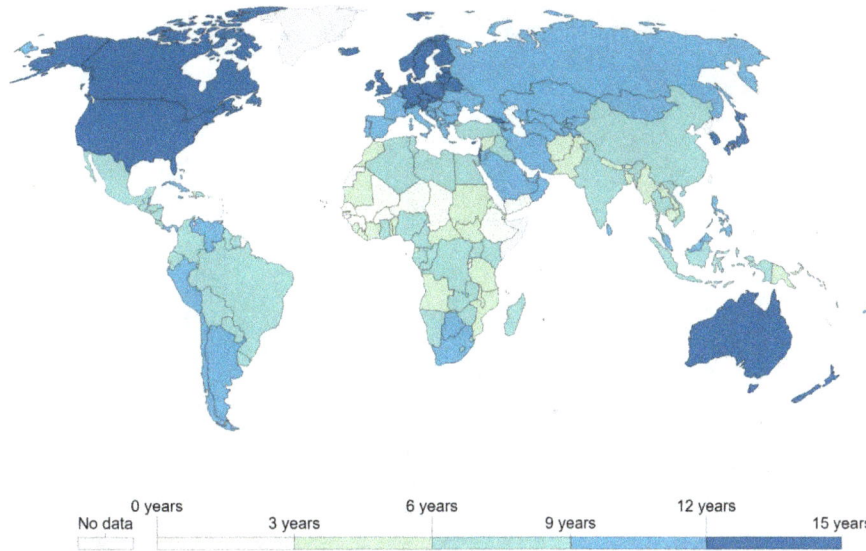

Mean years of schooling, 2017

Average number of years of total schooling across all education levels, for the population aged 25+

Source: Lee-Lee (2016), Barro-Lee (2018) and UNDP, HDR (2018)   OurWorldInData.org/global-rise-of-education • CC BY

Mean Years of Schooling (for the population aged 25 and older)[cxliii]

Share of the world population older than 15 years with at least basic education

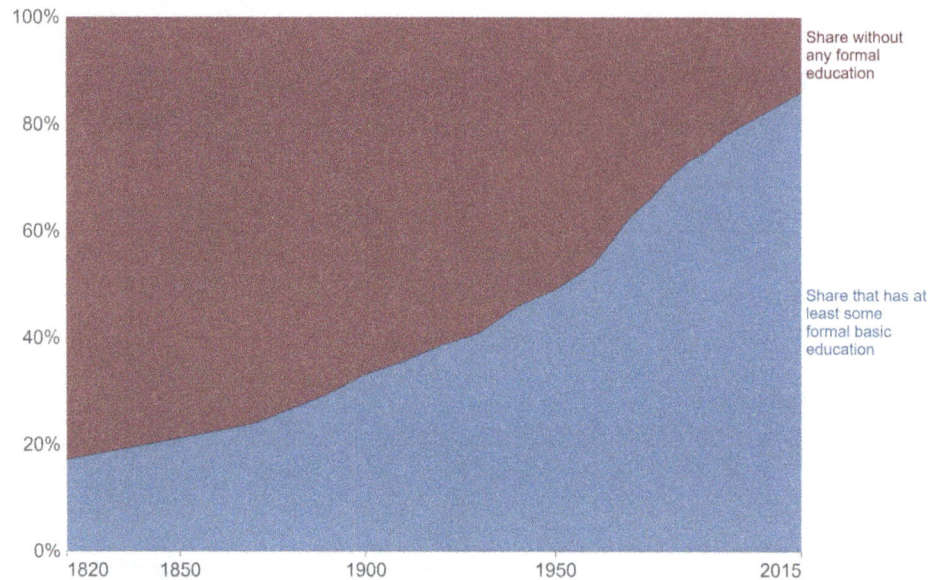

Source: Global education (OECD + IIASA (2016))   OurWorldInData.org/primary-and-secondary-education • CC BY

Share of the World Population Older Than 15 Years with at Least Basic Education[cxliv]

Life expectancy has increased from approximately 30 years in 1770 to over 60 years in most parts of the world. Child mortality (death before the age of 5) has also decreased from over 40% in 1800 to less than 5%. Through higher literacy and higher productivity and wealth, more time and resources can be used to apply and to communicate medical advances, thereby increasing life expectancy.

Life expectancy

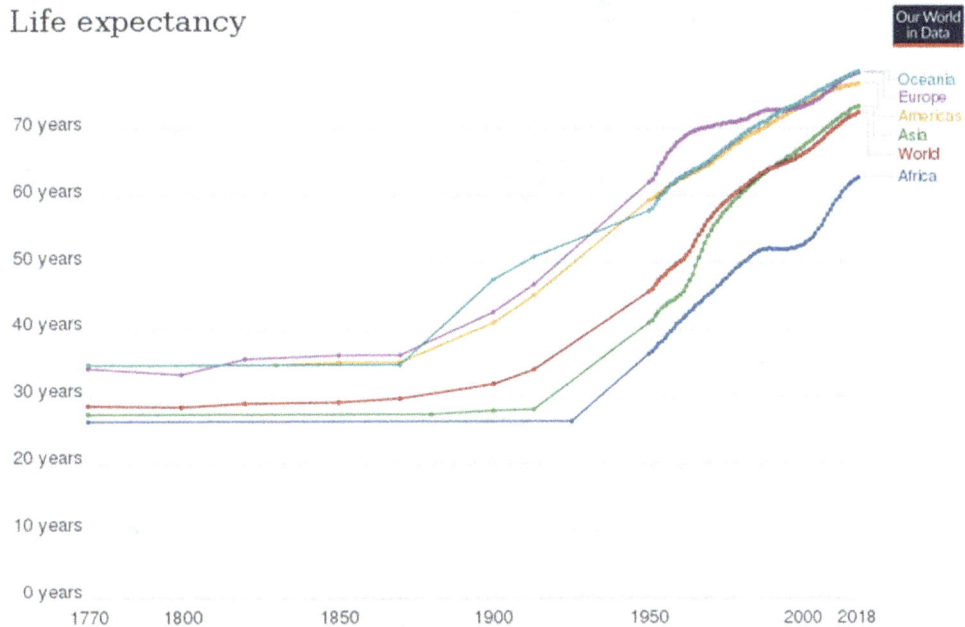

Legend: Oceania, Europe, Americas, Asia, World, Africa

X-axis: 1770, 1800, 1850, 1900, 1950, 2000, 2018

Y-axis: 0 years, 10 years, 20 years, 30 years, 40 years, 50 years, 60 years, 70 years

Source: Riley (2005), Clio Infra (2015), and UN Population Division (2019)
Note: Shown is period life expectancy at birth, the average number of years a newborn would live if the pattern of mortality in the given year were to stay the same throughout its life.

Life Expectancy[cxlv]

We have seen a change in occupations over history from hunter gathers, to farming and herding, to industrial work, and now to higher technology. In another feedback cycle, increased productivity allows for a greater diversity of work types, further increasing productivity.

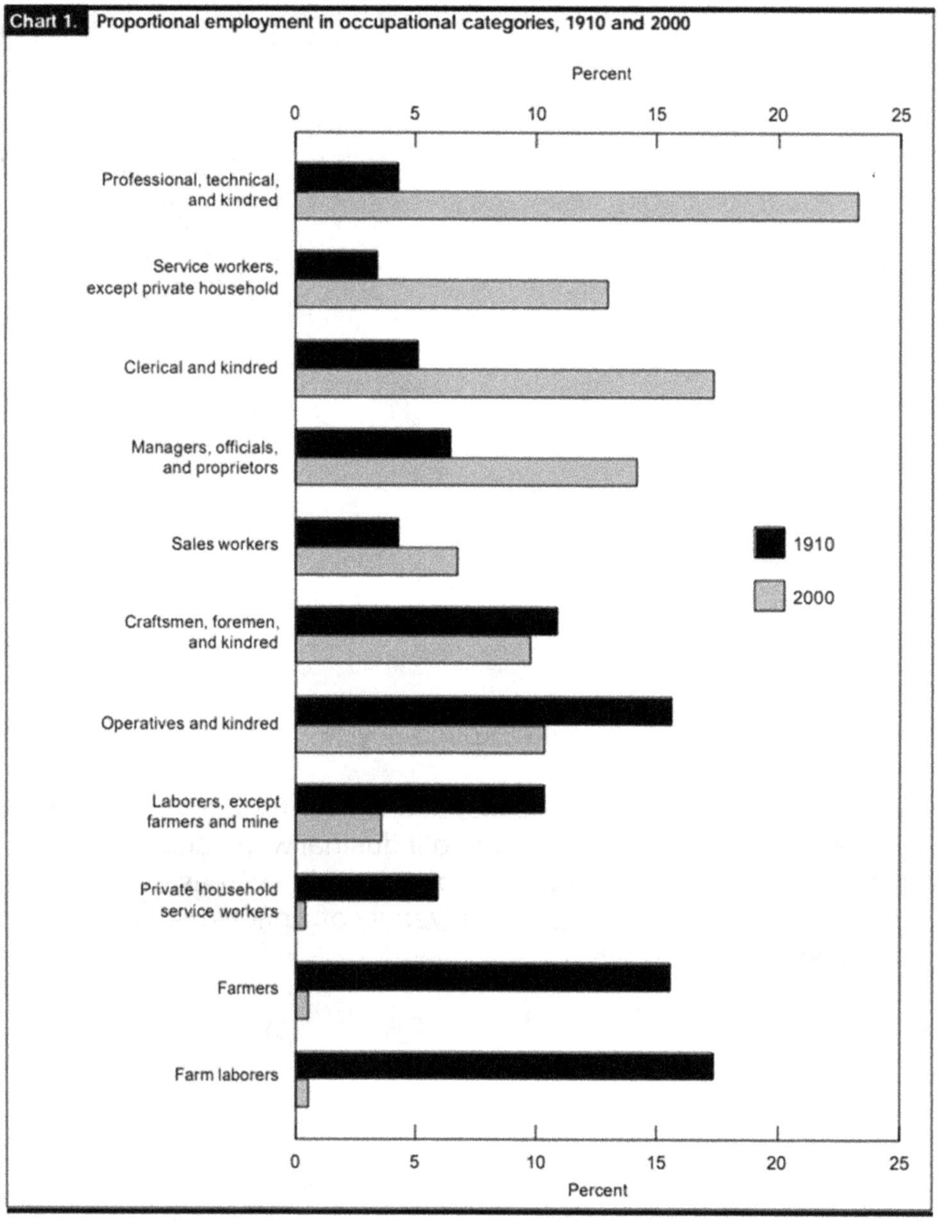
Chart 1.
Proportional employment in occupational categories, 1910 and 2000
Percent
0
5
10
15
20
25
Professional, technical, and kindred
Service workers, except private household
Clerical and kindred
Managers, officials, and proprietors
Sales workers
Craftsmen, foremen, and kindred
Operatives and kindred
Laborers, except farmers and mine
Private household service workers
Farmers
Farm laborers
1910
2000

Proportional Employment in Occupational Categories (1910-2000) – U.S. Data[cxlvi]

In many developed nations, routine labor roles are becoming obsolete. The University of Oxford estimates 47% of all U.S. jobs are at risk of computerization in the next 20 years. This trend is likely to spread to other nations as the standard of living increases in third world countries. While some jobs will become obsolete, it is anticipated that new types of jobs will emerge.

The U.S. workforce is more diverse by nearly every conceivable metric. Where once middle-aged, white males were the primary drivers of the labor force, we are now driven by a much less homogenous working population. Women, as of 2020 are the majority of the workforce[cxlvii]. We have an increase in the proportion of younger and older workers. The Pew Research Center shows that millennials became the largest generation in the labor force in 2015[cxlviii] (estimated to be 75% of the global workforce by 2025[cxlix]). At the other end of the spectrum, Americans ages 65 and older are projected as the fastest growing segment of the labor force.

Japan and Europe have aging populations with decreased birth rates. Immigration may provide some added personnel to the work force. This is more likely in Europe than in Japan due to cultural differences. Aging is creating more employment in medicine, health care, and servicing the needs of the elderly.

The work force will need to be able to learn new skills while meeting the demands of their current roles. Education will likely need to be more focused on the ability to learn new skills as technology changes accelerate. Memorization of facts and learning of specific skills will have limited utility as workers may need to reinvent themselves during a lifespan that may continue to increase as replacement organs and personalized medicine become commonplace. As automation and the use of artificial intelligence accelerate, microlearning[5] will be needed to allow flexibility to adapt to changing needs.

The electronic commerce has made transactions efficient, secure. Online stores can have low operating costs. Currently transitions are being made from using paper money to credit cards and digital

---

[5] Microlearning is skill-based learning that involves small learning units. It uses short-term, focused strategies designed for skill-based understanding.

wallets, due in part to e-commerce. The widespread use of electronic banking and credit has diminished the direct personnel roles in banking and credit unions. One of the most important features of digital payments is the fact that they are easily traceable. Adoption of newer technologies such as cryptocurrency or digital assets, created using computer networking software for transactions, may impact industries such as banking. These new distributed technologies will likely further diminish the personnel roles and increase the computerized roles in finance. Currently, most cryptocurrencies use blockchain technology to record transactions. A blockchain is a decentralized and distributed digital ledger consisting of records called blocks that are used to record transactions across many computers so that any involved block cannot be altered retroactively, without the alteration of all subsequent blocks. Also, many organizations are using artificial intelligence in digital commerce to improve customer satisfaction, revenue, and cost reduction.

Religious Changes

In the past, when peoples were much more isolated, district religions rose in different regions. Globalization started in the 1600's began to intermix regions, particularly in the newly settled Americas of the New World. After initial clashes (see my first book "The Love of God is a Root of Evil"), cross cultural influences of various non-religions and religions have led to greater tolerance and irreligiosity.

It appears in America that there is an increasing number of people that do not identify with a religious group[cl]. Per Pew Research, the religious composition in the United States will continue to show a drop in Christians and an increase in unaffiliated.

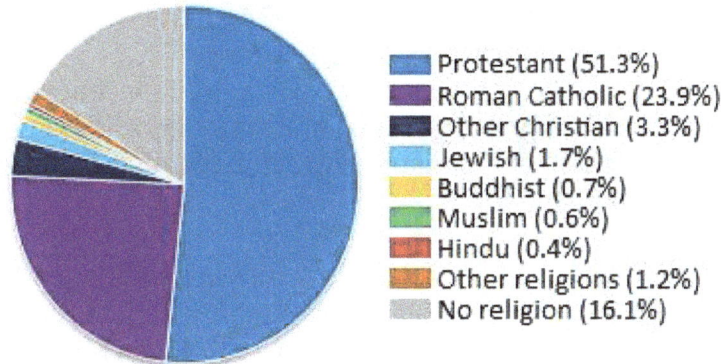

Religious Composition in the United States (2009)[cli]

According to Pew Research, many people who report "none" on religion state that they are scientists and do not believe in miracles. Others state that they do not believe in God because of common sense, logic, or a lack of evidence. Some state that they do not like the hierarchical nature of religious groups, think religion is too much like a business, or are concerned with clergy with sexual abuse scandals.

While those who live their lives according to secular values and humanist principles are on the rise, the percentage of the global population that are Muslims is expected to rise faster than any other religion. If current trends continue, the number of Muslims will approach that of the number of Christians around the world by 2050[clii]. The global numbers of Muslims, Christians, Hindus, and Jewish populations are expected to increase.

According to the Pew Research Center's latest estimates, there were over 1.17 billion non-religious people in the world in 2015, and that number is expected to increase to over 1.2 billion by the year 2060[cliii]. However, since other religious groups are projected to grow much faster (such as Muslims), the global share of religiously unaffiliated people is expected to fall from 16% to 13% of the global population over the same time period.

In Japan, about 70% of adults claimed to hold personal religious beliefs sixty years ago, but today, that figure is down to only about 20%. For the first time in British history, there are now more atheists and agnostics than believers in God[cliv]. Church attendance rates in the UK are less than 2% of the British attending church on any given Sunday. Nearly 70% of the Dutch are not affiliated with any religion. According to Pew Research's latest predictions, the growth of seculars will level off within a few decades, while Islam will continue to grow, becoming the world's largest religion by 2050.

Most of the currently popular religions are of ancient origin but many new religions are of recent origin (i.e., Sikhism (1469), Mormonism (1830), Wicca (1904), Scientology (1954))[clv]. It is possible that a new religion may become popular.

TIMELINE:

2000 BCE	1900 BCE	563 BCE	0	622 CE

Hinduism	Judaism	Buddhism	Christianity	Islam
	Abraham born (c. 1900 BCE)	Life of Buddha (563-483 BCE)	Life of Jesus (4 BCE-30 CE)	Life of Muhammed (570-632 CE)

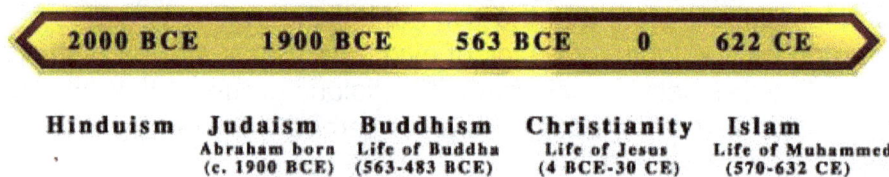

BCE – Before the Common Era
CE – Common Era

Timeline of Some Major Religions[clvi]

Political Changes

Over human history we have seen changes and variation in the types of government. Early government likely came from a family or clan structure. Tribal governments often had a king or chief that was often elected and had limited power. Many ancient kingdoms and empires had a king or emperor that was considered a god and therefore had to be obeyed.

Ancient Athens was a democracy, in that the adult male Athenian citizens voted directly on the affairs of state. With larger populations republics became practical, in which people elect other people to make decisions on their behalf. Elected officials in the Roman republic had to perform ceremonial religious duties as part of their jobs.

104

In the Middle Ages, Feudalism was a popular governmental structure due in part to the fall of the Roman Empire and the succeeding lack of safety for trade and travel. In Feudalism, the common people bound themselves to powerful rulers who offered security in exchange for labor and goods. European feudalism lasted until at least the 15th century.

During the Renaissance parliamentary government (king shares power with a group of officials) was used to help the king raise money by legitimizing the raising of taxes. The years around World War I largely saw the end of monarchy (hereditary rule by one person) as a legitimate form of government in many European countries, but monarchies still exist in Ghana, Nigeria, South Uganda, Indonesia, Malaysia, and the United Arab Emirates. Governments today remain diverse in their approaches to governing, but democratic-style government are common.

There are a few communist states including Cuba and North Korea. Many states remain dictatorships with limited political participation outside of an inner circle of rulers. Parliamentary government in a monarchy exists in Kuwait. There are also some theocratic governments, such the republic in Iran.

Some different types of government include:

Form of Government	Description of Governmental Form
anarchy	no order/control; no government structure; power vacuum
monarchy	rule of one; undivided rule; typically hereditary rule; backed by oligarchical power
oligarchy	rule of few (well-connected, socially, financially, physically powerful); elites rule
republic	indirect rule of citizens through representatives; rule of law; limited government
direct democracy	rule of citizens; simple majority rule; no restraint on majority
tyranny	micromanagement of citizens via government structure; military control, authoritarian
totalitarianism	total governmental control

Forms of Government[clvii]

Current trends towards a populism-driven electorate have been facilitated by a growing wealth inequality. Concentration of wealth has enabled the rich to have a greater influence the political system than others in society. In democracies relying on engagement with interest groups, decisions that tend to be unrepresentative of the public have increased disproportionately. Wealth inequality undermines education opportunities for children from poor socio-economic backgrounds, lowers social mobility and hampers skills development. Without redistribution of wealth, public unrest and increased crime may be expected. When social control in the community breaks down, perhaps due to low economic status or increased cultural heterogeneity, crime increases. Also, a perceived lack of possibility for social success can entice individuals to criminality. There are many factors involved the rate of public unrest and crime of which income inequality is one contributing factor (discussed further late).

U.S. Distribution of Wealth, 2007

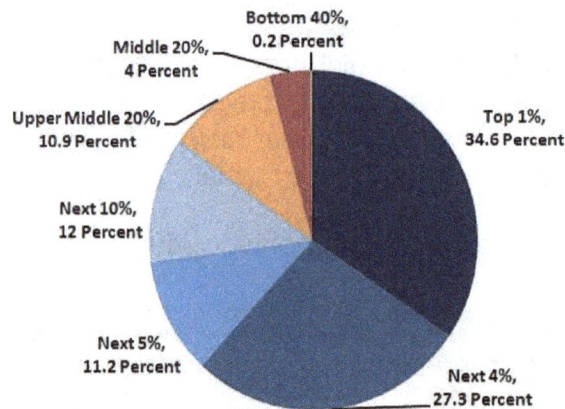

Bottom 40%, 0.2 Percent

Middle 20%, 4 Percent

Upper Middle 20%, 10.9 Percent

Next 10%, 12 Percent

Next 5%, 11.2 Percent

Next 4%, 27.3 Percent

Next 4%, 27.3 Percent

Top 1%, 34.6 Percent

Edward N. Wolff, 2010

U.S. Distribution of Wealth[clviii]

Today, slightly less than 1% of the world's adult population have over 1 million dollars. Despite their small numbers, this group collectively controls 46% of the world's wealth, valued at approximately $129 trillion[clix]. Not only is the money concentrated among a small portion of the world's population, those people tend to gravitate towards global cities such as London, Hong Kong, and New York. Seventy percent of persons with investable assets of $30 million or more reside in just ten cities around the world.

Humans are facing many challenges that require cooperation from many nations to allow success. Challenges associated with environmental pollution and global warming requires global support to decrease the impacts of the use of fossil fuels. As can be seen below, carbon dioxide emission is a problem that requires cooperation of many nations to decrease emissions.

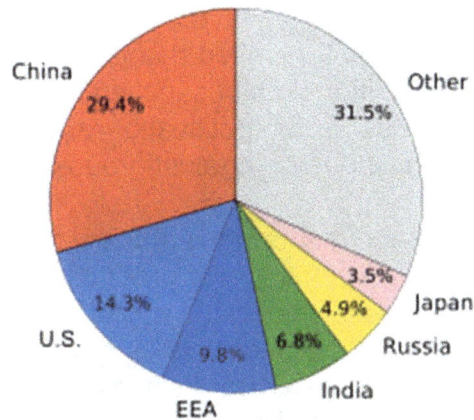

Global Carbon Dioxide Emissions by Jurisdiction
(EEA = European Economic Area)[clx]

Other global challenges include water shortages, terrorism, replacing energy sources, malnourishment, unemployment, and decreased biodiversity due to extinctions.

Organizations with worldwide influence are likely to take a prominent role in the future, while national governments will likely have a less important role. Many government services are likely to be provided by privately owned companies and mercenary forces.

Crime and Wars

Factors leading to increased crime rates, include crowed housing, economic depression, drug markets, and locations with major thoroughfares[clxi]. Some of the most troubled housing complexes are high-rise, family-occupied buildings concentrated in large cities, with tenants whose socioeconomic characteristics are linked to poverty. Increases in population may exacerbate these problems. While increased crime is likely to occur due to increased populations competing for limited resources and opportunities, it is unlikely to have a meaningful impact on the overall future existence of humankind. As technological advances develop in surveillance, such as the use of drones, videos, and internet usage, many crimes may be detected and punished effectively. Technological advances may also lead to more sophisticated types of crimes such as cyber-crimes, which are already common.

When there is competition for limited resources, warfare is likely to increase. Regional ambitions may be incited by the motivation to obtain desired resources. Some political or religious groups may use terrorist tactics to wage war against groups that they oppose.

In the current 21st century, shortages of fossil fuels, water or food will disrupt further the distribution and production of food itself, causing famines in regions vulnerable to global heating and drought. Already in many regions in Africa beyond the tropics, as well as the Middle East and Central Asia, high temperatures and scarce water and rainfall are already problematic for agriculture, so wars are likely to start in these regions as global heating increases. Such wars will further cause food shortages and societal disruption.

Humans have developed and continue to develop technologies for warfare. The use of nuclear, biological, chemical, and conventional warfare is likely to increase as nations compete for limited resources. The risk of biological and chemical warfare is going to increase. The availability of DNA editing systems like the "clustered regularly interspaced short palindromic repeats" (CRISPR) means that people with knowledge of how viruses work can develop a custom virus at minimal costs. CRISPR is a way of finding a specific bit of DNA inside a cell to be used in gene editing. The next step in CRISPR gene editing is usually to alter that piece of DNA. CRISPR has also been adapted to enable turning genes on or off without altering their sequence.

Wars are being fought remotely using drones striking enemies with precision. For years, the United States has deployed drones in the wars in Iraq and Afghanistan, and Turkish drones played a decisive role in fighting between Azerbaijan and Armenia in 2020[clxii].

Predator Drone with a Hellfire Missile[clxiii]

The drones are being enabled to use artificial intelligence. Drone technology itself is a relatively new area of military technology, but engineers combined drones with artificial intelligence to create a product that may be comparable to the performance of human reconnaissance teams[clxiv]. Usage of drones and robots in warfare will likely increase in the near future.

Other future means of destructive actions may involve the use of means of electronic bombs to disrupt communication grids as well high-energy laser weapons, space-based weapons, and nanotechnology (nanorobots for stealth missions). Some robotic warfare is already in use, and it is likely to increase in utilization by more technologically advanced nations.

It is interesting to note that the number of battle deaths per population shows a decrease in the last 80 years (see below). This is especially contrasting to the death rate that occurred during the time of World War II. This trend could easily be changed by the occurrence of a major global conflict.

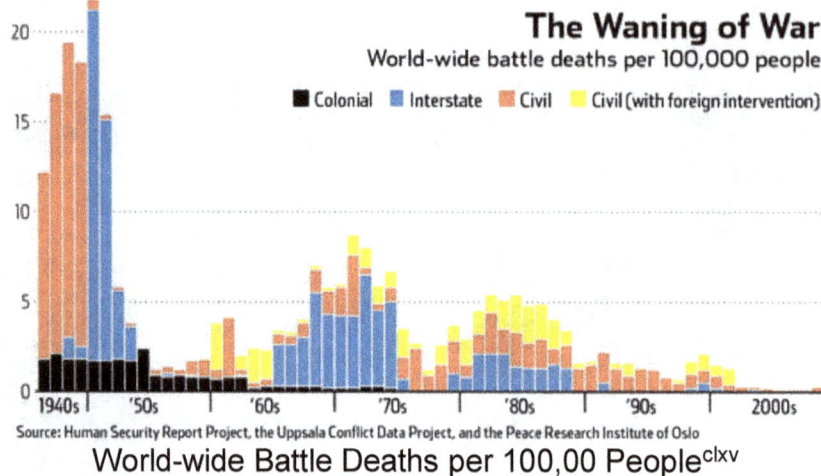

The Waning of War
World-wide battle deaths per 100,000 people

■ Colonial ■ Interstate ■ Civil ■ Civil (with foreign intervention)

Source: Human Security Report Project, the Uppsala Conflict Data Project, and the Peace Research Institute of Oslo

World-wide Battle Deaths per 100,00 People[clxv]

It is possible that a World War III scenario could develop under conditions associated with resource shortages due to overpopulation. While such an event is detrimental to society, it is unlikely that wars would lead to the annihilation of all humans. Global thermonuclear war is unlikely to eradicate every settlement on the Earth. Radiation and destruction would need to reach all areas, including the underground subways of major cities, the mountains of Tibet, the islands of the South Pacific, and Antarctica. Elaborate bunkers exist for government leaders to occupy during a nuclear war. Such events could lead to a massive loss of life not only to humans but to most life forms on Earth, but some humans would likely survive. The impact of such a war on society could be devastating.

Another potential problem associated with a global nuclear war is the potential for the blackening of the atmosphere with soot and dust, causing a reduction of sunlight. This could reduce global temperatures, which would reduce crop yields. Such additional consideration could cause massive numbers of deaths.

It is worthwhile to note that some forms of life are especially suited to survive in conditions where there are high levels of radioactivity. The red-colored bacterium *Deinococcus radiodurans* can resist 1.5 million rads of gamma radiation, about 3,000 times the amount that would kill a human[clxvi]. The bacterium survives and reproduces in environments that would be lethal for any other

110

organism and it also resists high doses of ultraviolet radiation. The most important component of this radiation resistance is the ability of the bacteria to repair damage to its chromosomal DNA. Other life forms that would likely survive a nuclear war include the amoeba, cockroach, scorpion, braconidae wasp, lingulate (marine invertebrate with an inarticulate shell and a long pedicle), fruit fly, mummichog (small marine killifish), and tardigrade (or water bear)[clxvii].

Such a global war could potentially set back society and cripple or slow advances in peaceful technological advances. As will be discussed later (see section on Colonization and Space Exploration), technological advances in space flight will be needed eventually to move Earthlings to other planets than Earth, if Earthlings are to survive in the long term. Avoidance or minimization of such a world war may be critical to our long-term survival. Currently, efforts are being made to develop rockets to reach and settle Mars cost-effectively (discussed further in the section on Colonization and Space Exploration).

Pandemics

Human history has seen many pandemics including the plague, black death, smallpox, cholera, Asian flu, Spanish flu, AIDS, and COVID. The largest known recorded numbers of human deaths were due to the Black Death (1347-1351 CE), Smallpox (1520 CE), Spanish Flu (1918-1919 CE), and Plague of Justinian (541-542) CE, though plagues have been reported even back in the Bronze Age, thousands of years before the Christian Era (BCE) (see the diagram on the next page).

HISTORY OF PANDEMICS

PAN-DEM-IC (of a disease) prevalent over a whole country or the world.

THROUGHOUT HISTORY, as humans spread across the world, infectious diseases have been a constant companion. Even in this modern era, outbreaks are nearly constant.

Here are some of history's most deadly pandemics, from the Antonine Plague to Novel Coronavirus (COVID-19).

Antonine Plague 165-180 5M
Plague of Justinian 541-542 30-50M
Japanese Smallpox Epidemic 735-737 1M
Black Death (Bubonic Plague) 200M 1347-1351
Small Pox 56M 1520
17th Century Great Plagues 3M 1600
18th Century Great Plagues 600K 1700
Cholera 6 outbreak 1M 1817-1923
The Third Plague 12M 1855
Yellow Fever 100-150K LATE 1800s
Spanish Flu 40-50M 1918-1919
Russian Flu 1M 1889-1890
HIV/AIDS 25-35M 1981-PRESENT
Asian Flu 1.1M 1957-1958
Hong Kong Flu 1M 1968-1970
SARS 770 2002-2003
Swine Flu 200K 2009-2010
MERS 850 2012-PRESENT
Ebola 11.3K 2014-2016
Novel Coronavirus (COVID-19) 6.4K* 2019-MAR 15 2020 [ON-GOING]

*As of Mar 15 officially a pandemic according to WHO

DEATH TOLL
[HIGHEST TO LOWEST]

200M
Black Death (Bubonic Plague)
1347-1351

56M
Small Pox
1520

40-50M
Spanish Flu
1918-1919

30-50M
Plague of Justinian
541-542

The plague originated in rats and spread to humans via infected fleas.

The outbreak wiped out 30-50% of Europe's population. It took more than 200 years for the continent's population to recover.

Smallpox killed an estimated 90% of Native Americans. In Europe during the 1800s, an estimated 400,000 people were being killed by smallpox annually. The first ever vaccine was created to ward off smallpox.

The death toll of this plague is still under debate as new evidence is uncovered, but many think it may have helped hasten the fall of the Roman Empire.

25-35M HIV/AIDS 1981-PRESENT
The Third Plague 1855
Antonine Plague 165-180
3M 17th Century Great Plagues 1665
Asian Flu 1957-1958
Russian Flu 1889-1890
Hong Kong Flu 1968-1970
Cholera 6 outbreak 1817-1923

A series of Cholera outbreaks spread around the world in the 1800s killing millions of people. There is no solid consensus on death tolls.

Japanese Smallpox Epidemic 735-737
600K 18th Century Great Plagues 1817-1923
200K Swine Flu 2009-2010
100-150K Yellow Fever LATE 1800s
11.3K Ebola 2014-2016
850 MERS 2012-PRESENT
770 SARS 2002-2003

6.4K* Novel Coronavirus (COVID-19) 2019-MAR 15 2020

VISUAL CAPITALIST

History of Pandemics[clxviii]

It is likely that we will continue to have more pandemics as there is an increase in population, intercontinental transport, and crowding. Some nations are involved in the development of offensive and defensive biological agents. Pandemics can be generated by natural mutations or by scientific development of biological warfare agents. The proliferation of travel increases the ease of transmission across geographic barriers. An engineered pathogen may be capable of wiping out all of humanity, if left unchecked. However, technologically advanced countries will likely be able to recognize and intervene to halt the spread of such a microbe and prevent human extinction[clxix]. Challenges may be encountered due to varied mutations of the bacteria or virus involved in the pandemic. Pandemics can temporarily decrease the population but are unlikely to have a lasting effect due to the development of vaccines and potential for gaining immunity after survival and population recovery.

While we are typically concerned with pandemics that involve death and illness to the human population, many animal and plant species have been decimated by the spreading of viruses, bacteria, and fungi. *Psudegymnoascus destructans*, white fungus, is responsible for killing bats in North America. The fungus *Batrachochytrium dendrobatidis* is responsible for massive frog death on a global scale that is threatening many species with extinction. With smaller gene pools associated with limited populations, it become easier to decimate plants and animal species, including vital agricultural plants and animals. The effects of a loss of biodiversity is discussed further elsewhere (see section on Extinctions).

Asteroids or Comets

As discussed previously, an asteroid of about six miles (10 km) in diameter struck the Earth approximately 65 Mya near the present-day Yucatan peninsula of Mexico. The impact caused catastrophic conditions across the entire planet, including thick clouds of dust and ash that caused global temperatures to decrease, contributing to the extinction of about 75 percent of marine and land animals on Earth at the time, including the dinosaurs[clxx].

While the Earth's atmosphere protects us from tiny debris, some small meteors hit the Earth, causing limited localized damage. The chances of a major asteroid (1-2 km or 0.6-0.2 miles in diameter) hitting the Earth is estimated to be about a one in 300,000 chance every year, which could cause regional damage.[clxxi] An object of more than 5 kilometers (3 miles) in diameter could cause mass extinctions. Even though the probability of a major event is small in terms of a single year, the probability of a major collision becomes inevitable over a long time.

It is possible that in the future humans will be able to develop technology that is able to predict such a collision and alter the path of an oncoming asteroid or comet to avoid impact. In the next 50 million years it is expected that the Earth will suffer at least one large collision. The consequences of such an impact will vary depending on the size and location of the impact: oceanic impacts will devastate coastlines through resulting tsunami waves, while land strikes will ignite fires and eject dust high into the air for long time periods.

Artist's Rendering of Asteroid Collision[clxxii]

Solar Storms

Another astronomical threat to human civilization might come from the Sun, which sometimes emits bursts of magnetized plasma into space. In a strong enough outburst, that in a rare event might strike the magnetic field of the Earth, the resulting electromagnetic storm on the surface can devastate electronic equipment, upon which most of the modern world depends for existence. Communications, transportation, farming, power distribution and a host of other vital functions of civilization might collapse or be severely impaired leading to data loss and a possible return to Dark Ages, or mass deaths.

Evolution

We have discussed earlier the origin of life and the various species that have evolved over the past billions of years. The past species and existing life forms have adapted to function in the environmental conditions through changes or mutations in genetic traits. If events had occurred differently (asteroid collision not occurred or at a different time) or mutations had occurred differently, life forms today might be significantly different. Although humans are vastly outnumbered by viruses and bacteria, and are less prominent than plant life, human activities are now causing great changes to the Earth's environment to rival those of bacteria.

Humans are causing global warming and other changes to the environment, as discussed previously. These changes are leading to a sixth massive extinction (Holocene or Anthropocene Extinction). Changes in environmental conditions and populations of organisms can lead to long-term evolution in the population characteristics. Some populations that can compete or cooperate to adapt to the changes are likely to survive, while those that cannot adapt will tend toward extinction. There is an increasing rate of environmental change, including habitat destruction due to human intervention. This loss is leading to the extinction of many species. While extinction is detrimental to those that are eliminated, it has made and will continue to make opportunities for new species to develop (for example, extinction of the dinosaurs enabled the rise of mammals).

Many of the endangered species are at the top of the food chain (often predators) whose numbers are dwindling due to conflicts with humans. Humans often kill predators for our own interests, and we destroy their habitats to expand our communities and agricultural operations[clxxiii].

Some species have been transported from their native areas to other regions. Some of these species have thrived in the new locations and are able to displace species that were native to the new location. Such introduction of "alien" species can often harm the biodiversity and disrupt the food chain.

Some examples of invasive species include European rabbits and Japanese beetles. European rabbits were introduced to Australia in the 18th century and eventually became widespread and numerous. Such wild rabbit populations are a serious mammalian pest in Australia, causing damage to crops. Although biological controls (myxoma virus and rabbit haemorrhagic disease virus) have provided ongoing rabbit control for over 60 years with little cost, they have not eliminated the problem. The overall loss caused by rabbits to agriculture and horticulture in Australia was recently estimated to be about $206 million per year[clxxiv].

The Japanese beetle is a destructive plant pest of foreign origin. It was first found in the U.S. in 1916 and has since spread to most states east of, and immediately to the west of, the Mississippi River. It has also spread to some western states, but regulations and careful monitoring have prevented its establishment elsewhere. The Japanese beetle has become a serious plant pest and a threat to American agriculture.

As species adapt to environments, the distribution of genetic traits can be altered. Certain mutations may be found to provide advantage to the mutated population over those without the mutation. Over time these modifications may lead to differentiation of species. New species may develop that are adaptive to the environmental changes caused by global warming, changes in the

116

levels of environmental pollutants, and the modification of land to agricultural and urban uses.

Extinctions

It is estimated that there about eight million species on Earth. Of these, at least 15,000 are currently threatened with extinction[clxxv]. At the end of the last ice age, 10,000 years ago, many animals went extinct, including mammoths, mastodons, and glyptodonts (heavily armored armadillos).

Glyptodont Fossil, Natural History Museum, Vienna[clxxvi]

Beginning in 1800, industrialization drove up the extinction rates and has continued to do so (see graph). Extinctions are now running at 100 to 1,000 times the background rate.

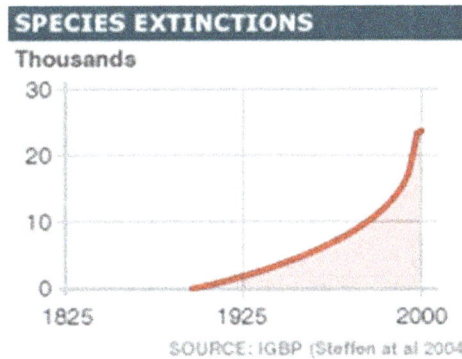

SPECIES EXTINCTIONS
Thousands

SOURCE: IGBP (Steffen at al 2004)

Estimate of Recent Species Extinctions
(Extinctions in Thousands versus Time)[clxxvii]

With increased human populations, land usage, water usage, and overhunting and overfishing, many species have become extinct and many more are expected to become extinct. For a detailed discussion of the impact of humans on life on Earth, see *"The*

117

Future of Life" by Edward Wilson[clxxviii]. The International Union for Conservation (IUCN) provides a Red List of threatened species[clxxix]. The Red List is an indicator of the health of the world's biodiversity. The table below provides a summary of this information.

	Estimated Number of described species[?]	Number of species evaluated	Number of threatened species[?]	Number threatened, as % of species described[?]	Number threatened, as % of species evaluated[?]
Vertebrates					
Mammals[1]	5,488	5,488	1,141	21%	21%
Birds	9,990	9,990	1,222	12%	12%
Reptiles	8,734	1,385	423	5%	31%
Amphibians[2]	6,347	6,260	1,905	30%	30%
Fishes	30,700	3,481	1,275	4%	37%
Subtotal	**61,259**	**26,604**	**5,966**	**10%**	**22%**
Invertebrates					
Insects	950,000	1,259	626	0%	50%
Molluscs	81,000	2,212	978	1%	44%
Crustaceans	40,000	1,735	606	2%	35%
Corals	2,175	856	235	11%	27%
Arachnids	98,000	32	18	0%	56%
Velvet Worms	165	11	9	5%	82%
Horseshoe Crabs	4	4	0	0%	0%
Others	61,040	52	24	0%	46%
Subtotal	**1,232,384**	**6,161**	**2,496**	**0.20%**	**41%**
Plants[3]					
Mosses[4]	16,000	95	82	1%	86%
Ferns and allies[5]	12,838	211	139	1%	66%
Gymnosperms	980	910	323	33%	35%
Dicotyledons	199,350	9,624	7,122	4%	74%
Monocotyledons	59,300	1,155	782	1%	68%
Green Algae[6]	3,962	2	0	0%	0%
Red Algae[6]	6,076	58	9	0%	16%
Subtotal	**298,506**	**12,055**	**8,457**	**3%**	**70%**
Others					
Lichens	17,000	2	2	0%	100%
Mushrooms	30,000	1	1	0%	100%
Brown Algae[6]	3,040	15	6	0%	40%
Subtotal	**50,040**	**18**	**9**	**0.02%**	**50%**
TOTAL	**1,642,189**	**44,838**	**16,928**	**1%**	**38%**

Summary of Threatened Species (IUCN Red List)[clxxx]

More vertebrates, which have been intensively described, have been evaluated for extinction more than other life forms. 76 vertebrate species have become extinct since the year 1500 CE. Overall, 22% of the evaluated vertebrate species are known to be threatened or extinct.

The invertebrates are estimated to make up 97% of all species (including many insect species). A much smaller proportion of invertebrate species have been evaluated, but of those, 41% are threatened or extinct.

Reducing biodiversity will impact societies at several levels, including diminishing the availability of economically valuable natural goods such as timber and compromising ecosystem functions such as fresh water and biodegrading bacteria. Life forms on Earth are interconnected and dependent on each other. Impacts on one species has impacts on other species. If you have extinction of some species that are food sources for another species, the dependent species may have problems in obtaining nutrition. Also, there are competitive balances between species and symbiotic relationships that may be disrupted.

For example, human intervention of the gray wolf population had subsequent effects due to their dwindling population numbers. Before a mass extermination effort in the U.S. that decreased wolf populations in the first half of the 20th century, wolves kept other animals' populations from growing exponentially. They hunted elk, deer, and moose and killed smaller animals such as coyotes, raccoons, and beavers. Without wolves to keep other animals' numbers in check, prey populations grew larger. Exploding elk populations in the western U.S. wiped out so many willows and other plants that songbirds no longer had sufficient food or cover in these areas, threatening their survival and increasing numbers of insects like mosquitos that the songbirds control.

The losses of large species like the wolf, tiger, rhinoceros, and polar bear are better known than the disappearance of moths or mussels, which are small species that can affect ecosystems in significant ways. There are nearly 300 species of mussels in North American river and lakes, and most of them are threatened. Mussels play an important role in the aquatic ecosystem. Many kinds of wildlife eat mussels, including raccoon, otters, herons, and egrets. Mussels filter water for food and thus are a purification system. They are usually present in groups called beds. Beds of mussels may range in size from smaller than a square foot to many acres; these mussel beds can be a hard cobble on the lake, river, or stream bottom, which supports other species of fish, aquatic insects, and worms. In their absence, these dependent species settle elsewhere, lower the available food source for their predators, and cause those predators to leave the area.

There currently are efforts to save species of various animals by environmentalist organizations through protecting, restoring, and connecting the habitats on which endangered species and other wildlife depend for their survival, and encouraging wildlife-friendly land management. They are also trying to reduce threats to wildlife that can lead to their endangerment and extinction, such as loss of habitat, contamination of water, and spread of invasive species. It does not appear that such efforts will be able to counteract the current Anthropocene extinction, as evidenced by the IUCN Red List of Threatened Species[clxxxi] that grows every year.

Potential Scientific Advances

We have seen that technological advances have accelerated over history. Humans have progressed from stone tools (3 Mya), copper smelting (5000 BCE), iron smelting (1300 BCE), windmill (650 AD), printing press (1439 in Europe, earlier in China), steamboat (1783), first gasoline power automobile (1886), liquid fueled rocket (1926), first microprocessor (1971), World Wide Web (1990), and the first solar sail-based spacecraft (2010). Inventions and technological advances are expected to continue as finances and manpower are expended on academic and industrial research efforts.

Some scientific advances are more predictable due to knowledge of current research efforts and understanding of the technological hurdles that need to be overcome. Scientific advances in the far future are much more difficult to predict or even imagine. Science fiction authors sometimes imagine technologies that may be possible in the future but some of these are likely to remain fiction due to the apparent scientific impossibilities or impracticality of bringing these technologies to reality. Advances such as transporters (as in Star Trek), use of wormholes for space jumps and time travel, and warp speed are unlikely based on our current scientific understanding. Yet, our understanding of areas such as dark matter and quantum physics are limited, and better knowledge of these areas of physics may lead to technologies that are not currently imagined.

Humans have and continue to make technological advances that will impact the future. Advances in agriculture, energy production, information storage and processing, manufacturing, medicine, and space exploration may aid in the long-term future survival of humans. Advances in widely applicable technologies are already having an impact. Some of these widely applicable technologies include increasing computer capability, artificial intelligence, nanotechnology, biotechnology, and three-dimensional printing. The use of devices for surveillance are being used in many applications from security, warfare, medicine/health monitoring, finances, marketing, etc.[clxxxii] Surveillance technology will likely limit privacy in exchange for greater efficiency and increased knowledge, but can also stifle innovation.

Some are concerned that wide applications of artificial intelligence will potentially lead to loss of jobs. This type of concern, associated with job loss due to technology advances, has been repeatedly expressed over history. Past advances in technology have led to the end of the need for some occupations and the advances have led to the development of new jobs. For example, stage wagon drivers are no longer needed but the need for cab and truck drivers has developed because of the development of the automobile, which replaced the stage wagon. The development of drone technology has generated the need for new jobs associated with the operation of drones, while the need for some soldiers/airmen may have decreased for some missions.

Advances in computers, artificial intelligence, nanotechnology, robotics, biotechnology, and three-dimensional printing are likely to change the types of jobs that humans perform. While computing power has increased over recent history, it is anticipated that quantum computers will be able to quickly solve certain problems that no classical computer could solve in any feasible amount of time. The most widely used type of quantum computer is based on the quantum bit, or "qubit", which is somewhat analogous to the bit in classical computation. A qubit can be in a 1 or 0 quantum state, or in a superposition of the 1 and 0 states. When it is measured, however, it is always 0 or 1. The probability of either outcome depends on the qubit's quantum state immediately prior to measurement. A notable application of quantum computation is for attacks on cryptographic systems that are currently in use. Also, this technology may be helpful in applications such as drug discovery.

The use of artificial intelligence can be applied to a wide variety of applications including autonomously driving transportation, algorithm development, and robotics. Autonomously driving cars and trucks will displace the need for drivers but is likely to open other types of jobs associated with the refinement of this technology and networking traffic systems with autonomous vehicles. Autonomous driving transportation is likely to improve the safety of human transport. Perhaps further in the future autonomous flight of vertical takeoff and landing vehicles could be developed and used to improve transportation. Artificial intelligence can be used to develop algorithms by training from large data sets. Artificial intelligence is being used for pattern recognition to detect and recognize objects, which is needed for many robotic processes. Some such as Ray Kurzweil believe that advances in artificial intelligence will increase exponentially and that we will be able to multiply our effective intelligence a billion-fold by merging with the intelligence we have created[clxxxiii].

Nanotechnology, which can be defined as the manipulation of matter with at least one dimension sized from 1 to 100 nanometers, applications will continue to expand. Nanotechnology is currently used in areas of including: information technology, homeland security, medicine, transportation, energy, food safety, and environmental science. Nanoscale additives to or surface treatments of fabrics can provide lightweight ballistic energy deflection in personal body armor, or can help them resist wrinkling, staining, and bacterial growth. Researchers are investigating carbon nanotube "scrubbers" and membranes to separate carbon dioxide from power plant exhaust. Nanotechnology could help meet the need for affordable, clean drinking water through rapid, low-cost detection and treatment of impurities in water. Nanotechnology offers the promise of developing multifunctional materials that will contribute to building

122

and maintaining lighter, safer, and more efficient vehicles, aircraft, spacecraft, and ships.

Three-dimensional (3D) printing is currently being used in the areas including manufacturing (customization, rapid prototyping, and food production), medical applications, apparel/jewelry, automotive parts, and home construction. Commercially available printers are currently available for limited applications (see photo on next page). 3D printing will likely decrease the need for transport of some manufactured goods by local manufacturing at printer sites.

Creality 3D Printer[clxxxiv]

Agricultural Technologies

Increased population and improved standards of living will require increased food production. Advanced technologies have been and will continue to be developed to address the increased need for food.

The use of sensors for detecting soil and plant conditions in combination with data on present and predicted weather conditions would allow artificial intelligence systems to recommend or make modifications for optimization of plant growth and health in real time.

The use of plant-based protein and stem cell-based muscle tissue grown in a laboratory will potentially reduce the negative impacts of livestock farming on land, water, and greenhouse effects. With the global population expected to increase to approximately 9.7 billion by 2050, meat alternatives could be effective in creating a more sustainable food supply without forcing people to change their diet drastically[clxxxv].

The use of robotic tractors, combines, grain trucks, and sprayers would allow for reduction in labor costs and potentially improved productivity.

Robotic Farming[clxxxvi]

Genetically modified crops and livestock have increased productivity and can potentially have greater increases in productivity. In 2015, 92% of corn, 94% of soybeans, and 94% of cotton produced in the U.S. were genetically modified varieties. Some express concern about the potential negative effects of consumption of genetically modified foods, such as food allergies, liver problems, reduced nutritional value, increased toxicity, and antibiotic resistance. These concerns appear to be equally prevalent in conventional food[clxxxvii]. The scientific research conducted to date has not detected any significant hazards directly connected with the use of genetically engineered crops[clxxxviii]. Concerns associated with negative impacts of genetically modified food will need to be monitored. Genetically modified food has the potential to improve fruit and vegetable shelf-life, improve nutritional value, and improved quantity and quality of meat, milk, and livestock. Desired genetic characteristics of the crops and livestock can be chosen to produce desired effects.

Indoor agriculture using hydroponics and aquaponics may become more widespread in use, potentially leading to better land use and optimization.

Automated Hydroponics[clxxxix]

Some predict that people are likely to become vegetarians in the future. Artificial meats, which are being developed, are likely to be cheaper, tastier, and healthier than meat from slaughtering animals. Eating of real animals may become considered barbaric in the future. This would lead to less breeding of animals for the meat industry with potential benefits to environment (decreased water usage, less deforestation, lower Greenhouse emissions, etc.). The use of three-dimensional printing may help enable the generation of creative and preferred food products.

Advances in technology for food production could address concerns associated with water and land usage as well as concern associated with sustaining the ability to feed billions of people. A decrease in population would make these challenges easier to address.

Energy Production

The energy needs of humans has increased significantly with the increase in population and with the improvement in standards of living. Movement from the of use of fossil fuels such as oil, coal, and natural gases is inevitable as the sources of fossil fuels are depleted and the cost of obtaining the resources increases. Some technological advances have helped provide access to additional fossil fuels such as fracking, which injects liquid at high pressure into subterranean rocks to extract oil or gas. Currently, the world consumes about 14 trillion watts of power, of which 33 percent comes from oil, 25 percent from coal, and 20 percent from natural gas. The negative impact of using these resources on the environment has already been discussed.

Alternative energy sources are being used and will likely continue to become more prominent. Wind, geothermal, solar, biomass, hydroelectric, and nuclear (fission reactors) power are some types of energy being used. Currently, about 15 percent of our energy comes from biomass and hydroelectric, 7 percent from nuclear, and 0.5 from solar and renewables. The use of hybrid and electric cars (likely to become fully autonomous) will reduce the need for gasoline but will increase the need for electrical power. Currently the generation of energy using biomass can lead to deforestation (wood is a significant source of biomass energy) and harmful emissions including lead, nitrogen oxide, particulate matter, carbon monoxide, and sulfur dioxide.

In the future some advanced technologies are likely to become available to address the expected increased energy needs. Some of these technologies are discussed below.

Thorium-based reactors could provide an improvement over uranium-based fission reactors. The thorium fuel cycle offers several potential advantages over an uranium fuel cycle, including thorium's greater abundance, better physical and nuclear properties (e.g. lower capture-to-fission ratio for thermal neutrons), reduced plutonium and actinide pollutants, and better resistance to nuclear weapons proliferation when used in traditional light water reactors[cxc]. France, India, Japan, Norway, China, and the United States are working on thorium nuclear reactors[cxci]. China is working toward development of a molten salt reactor that uses thorium as fuel by 2030.

Research on the generation of power from nuclear fusion has been and continues to be leading toward a potential solution for providing vast amounts of energy. Nuclear fusion has a larger energy generation to fuel ratio than nuclear fission. Unlike nuclear fission, nuclear fusion does not generate long-lived nuclear waste products[cxcii] and would be a safer technology (reactors cannot melt down, the reactor can be stopped instantly at any time, the nuclear waste cannot be used in bombs)[cxciii]. In southern France, 35 nations are collaborating to build the world's largest tokamak, a magnetic fusion device that has been designed to prove the feasibility of fusion as a large-scale and carbon-free source of energy. This project is called the International Thermonuclear Experimental Reactor (ITER). ITER is expected to be the first fusion device to produce net energy.

ITER Facility in 2021[cxciv]

Other approaches to development of fusion technology include the FuZE-Q reactor, which bypasses the need for costly and complex magnetic coils, as in the case of tokamaks. Instead, the machine sends pulses of electric current along a column of highly conductive plasma, creating a magnetic field that confines, compresses, and heats the ionized gas[cxcv].

The use of superconductors may enable reduction in energy loss transmission, increasing energy efficiency. This may also enable the use of magnetic fields for magnetic trains and magnetic cars. The use of room-temperature superconductors could enable the use of flying cars and trains that would float on rails over superconducting pavement without friction[cxcvi]. Improved battery technology could lead to proliferation of electric aircraft.

The use of space satellites to absorb radiation from the Sun and beaming this energy to the Earth may provide a new source of energy. The current problem with using space solar power is the cost of launching the space collectors.

Some other technologies under research include:
- Geothermal Energy - Heat from below the surface of the Earth, piped up as hot liquid or transferred in heat pumps
- Wastewater Electricity Generator – uses microbial fuel cells and reverse electrodialysis to use wastewater to produce electricity
- Waste-Sourced Biofuel – turns biomass waste into gas and ethanol and burns agricultural residue through pyrolysis (thermal decomposition of materials at elevated temperatures) into charcoal
- Solar Glass – photovoltaic glass that can convert sunlight into electricity, which could be put into the windows of offices and homes
- Graphene – Strong, flexible, conductive single layer graphite suitable for transfer of energy over distance with minimal loss
- Molten Salt Storage – use of molten salt for storing energy for future use when used in solar energy production
- Smart Grids/Metering – enabling more efficient power conservation, programming device use during off-peak energy demand hours, smoothing out demand
- Artificial Photosynthesis – use sunlight and carbon dioxide to produce energy, combination of carbon capture and conversion of solar energy into electricity
- Green Architecture – buildings to be constructed using natural light to reduce the need for lighting and with adequate insulation to eliminate the need for heating

Technological developments in energy production could help address the concern over competition for limited energy resources, which could lead to conflicts and wars. An abundance of energy could lead to a higher standard of living. The energy production would ideally be provided by sources like the Sun or matter conversion with limited generation of pollution.

Manufacturing

Robots have been used in manufacturing for decades. Robots are currently used in manufacturing to take on repetitive tasks, which streamlines the overall assembly workflow. Robots have the advantages of not needing rest to avoid making mistakes, can minimize inconsistency, and can be safely used in dangerous tasks. Efforts are in progress to increase the use of human robot collaboration using voice recognition, natural language and gesture understanding, and the use of mobile robots in manufacturing using autonomous navigation, mobile manipulation, and sensor fusion. The proliferation of robotic technology may lead to the use of robots in the construction of buildings.

Industrial Robot Used in Manufacturing[cxcvii]

Improvements in manufacturing may use biomimicry to improve structural and material properties. Biomimicry in architecture and manufacturing means designing buildings and products to mimic naturally occurring processes. There are ultra-strong synthetic spider silks, adhesives modeled after gecko feet, and wind-turbine

129

blades that mimic whale fins. There are currently efforts to build screen systems that use elasticity, geometry, and thermobimetal to open and close in response to sunlight, like a flower[cxcviii]. Biomimicry may lead to production of self-healing material that would enable longer lasting consumer goods.

More than eight trillion kilograms of plastic have been produced to date, and eight billion kilograms of plastic flow into the ocean every year. It ensnares marine animals and fish. It appears in the table salt we use, and it is found in our bodies[cxcix]. Efforts toward the use of plant-based plastic (instead of petroleum based) that would be biodegradable are being investigated. In the United Kingdom, one boutique is growing fungus into lightweight furniture, and in the U.S. the Department of Agriculture is using a milk film to create packaging that keeps food fresh[cc]. In the future, products may be made from biodegradable materials, which may help with problems of pollution and waste production.

The use of three-dimensional printing may enable local manufacturing of goods as opposed to using inexpensive labor in remote locations and then transporting the goods. As three-dimensional printing evolves, more complex designs will be able to be made using various materials. Using multiple sources of material such as metals, cement, plastics, artificial wood, and biodegradable products will enable many products to be made. Printers that can alternate among multiple material sources could lead to production of complex products. Desired products can be manufactured using recyclable materials, which would lead to less waste and pollution. Large items such as housing can be printed using three-dimensional printers that are able to climb over the building being made.

Medicine

There has been considerable research conducted towards improving medicine over human history. One of the major improvements has been in decreasing the incidence and impact of infections due to a better understanding of bacteria and viruses. This has led to the use of vaccines and antibiotics, which has contributed to an increase in human life span. Other important

130

medical advances include the use of anesthesia, blood transfusions, insulin treatment[cci], dialysis, open heart surgery, transplant surgery, various advanced imaging methods, improvements in reproductive health (including birth control, c-sections, and tampons), and cancer treatments.

Our improved understanding of genetics is likely to lead to improvements in precision medicine. Precision medicine will involve using an individual's genetic information to allow use of treatments that are preferentially effective for certain genotypes. Obtaining and using individual genetic information could also lead to treatment of genetic disorders using gene therapy[ccii].

A wide range of robots are being developed to serve in a variety of roles within the medical environment. Robots specializing in human treatment include surgical robots and rehabilitation robots[cciii]. The field of assistive and therapeutic robotic devices is also expanding. Robotics are being used to help examine and treat patients in rural or remote environments. The integration of robotics and telemedicine may lead to tele-nursing, where a human nurse can remotely control a robot to perform many of the tasks involved in patient care[cciv]. The proliferation of robotic technology may lead to the use of small robotic devices that could be used in medicine for minimally invasive localized treatment.

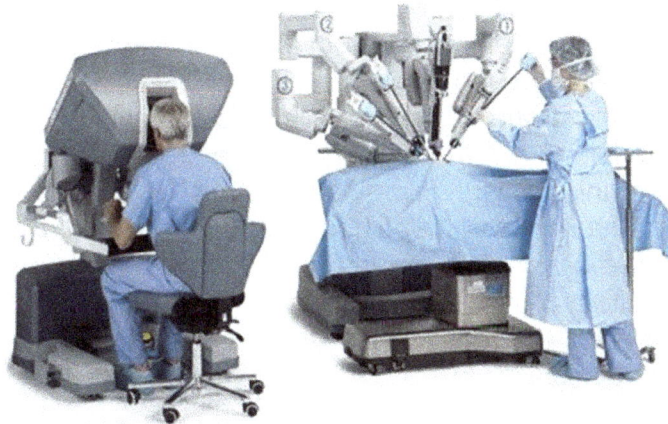

Current Surgical Robotic System[ccv]

The bioprinting of various organs could enable replacement of damaged or diseased organs. This could be achieved using three-dimensional printing for generation of replacement organs. Stem cells or derivatives could potentially be used to form these organs[ccvi].

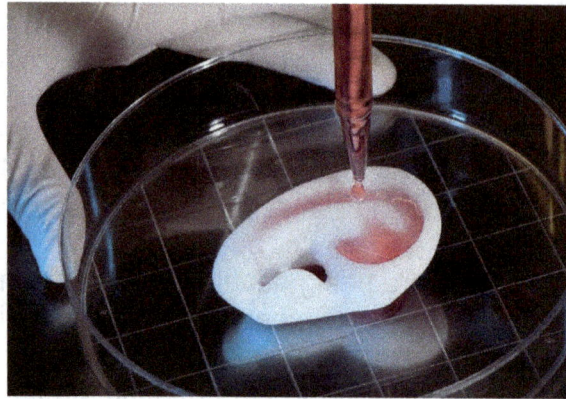

Three-Dimensional Printed Ear Lobe[ccvii]

The development of replacement organs could lead to extension of life for many people who would otherwise die. Currently, many people die on waiting lists for organs. If the cells used in development of these organs can be made to have low to no rejection, this could lead to improved quality of life.

Research to address the effects of aging is ongoing. Telomeres are highly repetitive DNA sequences located at the end of chromosomes, and telomere length is associated with cell age. As cells divide, telomere length gets progressively shorter until eventually, proliferation stops entirely. Technological advancements may lead to addressing the problem of telomere shortening during cell, ultimately decreasing the effects of aging. While some have hopes that ultimately such technology could lead to the elimination of aging and death, this currently seems unlikely. The use of stem cells in adult human tissues has the potential for cell replacement or tissue repair therapy in many degenerative diseases of aging[ccviii].

Colonization and Space Exploration

Ultimately, space colonization appears to be necessary for survival of Earthlings. This would enable some of the Earth's lifeforms to avoid any planetary-scale disaster (natural or man-made), and could provide additional resources to the Earth. Due to the eventual changes in our Sun and possible catastrophic events such as an asteroid collision with the Earth, space colonization is necessary for saving humans or their descendants (Earthlings).

In 2001, Steven Hawking, a theoretical physicist and cosmologist who made major contributions to the field of general relativity, predicted that humans would become extinct within the next thousand years, unless colonies could be established in space. He stated that humanity faces two options: either we colonize space within the next two hundred years and build residential units on other planets, or we will face the prospect of extinction.

The development of a space elevator (see diagram below) could serve to make space exploration more economically feasible. The space elevator is conceptually an Earth to space (approximately 36,000 km above the Earth) transportation system using a tether connecting the Earth to a satellite in space. Efforts are being made to develop the technology for a geostationary satellite with a tether for transporting materials and possibly humans from the Earth to space. Nanotechnology is hoped to provide a lightweight but strong material that could be used to provide the tether. Currently, there is no material that could provide the material properties needed for the tether. If carbon nanotubes without defects can be made into long cables or another lightweight, strong material can be produced, this would make the space elevator more feasible.

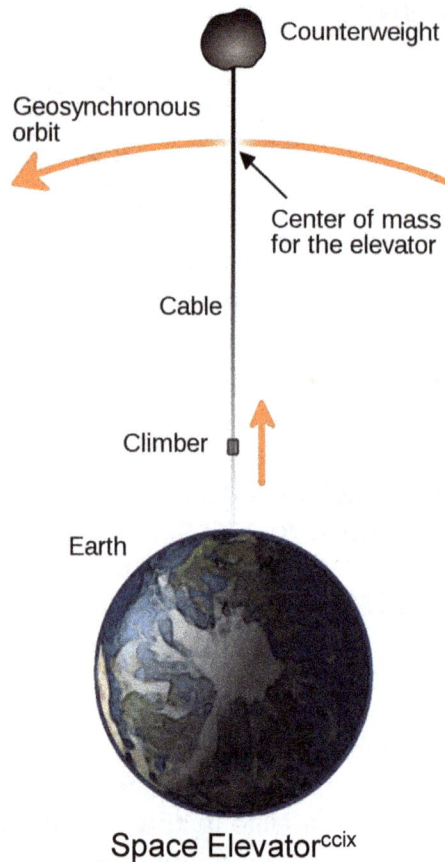

Space Elevator[ccix]

With or without the space elevator, efforts are underway to allow asteroid mining, which could provide materials for manufacturing in space. Asteroid composition varies from volatile-rich bodies to metallic bodies with high concentrations of rare metals such as gold, silver, and platinum as well as common elements such as iron and nickel. Platinum-rich asteroids may contain up to 100 grams per ton, which is 10 to 20 times higher than the open pit platinum mines in South Africa[ccx]. A 500-meter-wide platinum-rich asteroid could contain nearly 175 times the annual global platinum output, or 1.5 times the known world reserves of platinum group metals. The metals and other resources may be able to be used for construction in space or for return to Earth.

There are plans for Mars missions and colonization. Elon Musk (entrepreneur and business magnate) plans through the company SpaceX[ccxi], which is developing huge, cheap rockets, to establish a crewed base on Mars as early as 2050. Some technological challenges to overcome for success include:[ccxii]

- The health threat from cosmic rays and other ionizing radiation. In May 2013, NASA scientists reported that a possible mission to Mars may involve radiation risk based on energetic particle radiation measured by the radiation assessment detector on the Mars Science Laboratory while traveling from the Earth to Mars. The calculated radiation dose was 0.66 sieverts round-trip. The agency's career radiation limit for astronauts is 1 sievert.
- The adverse health effects of prolonged weightlessness, including bone mass, muscle mass, and eyesight impairment. In 2019, researchers reported that astronauts experienced blood flow and clot problems while on board the International Space Station, based on a six-month study of 11 healthy astronauts. These results may influence long-term spaceflight, including a mission to the planet Mars.
- Psychological effects of isolation from Earth and community due to lack of a real-time connection with Earth.
- Social effects of several humans living under cramped conditions for several years, depending on spacecraft and mission design.
- Lack of medical facilities.
- Potential failure of propulsion or life-support equipment.
- The complexity in establishing sustainable artificial ecosystems,

Building of a space colony presents a set of technological and economic challenges. Space settlements would have to provide for nearly all the material needs of hundreds or thousands of humans, in an environment that is hostile to human life. They would need technologies such as controlled life support systems. A space colony would be an expensive project. The use of robots could aid in the mining of materials and manufacture of habitats. The use of three-dimensional printing may allow for the manufacturing of needed housing and devices using materials derived from the planet. Replacement parts for devices as they degrade with use would not need to be stocked but instead could be made quickly on site.

Conceptual Mars Colony[ccxiii]

It is possible that by increasing greenhouse gases in the Martian atmosphere one could increase global temperatures. Once the temperature starts to rise, the underground permafrost may begin to thaw out. This could eventually lead to possible terraforming of Mars.

Building colonies in space will require access to water, food, construction materials, energy, transportation, communications, life support, simulated gravity, radiation protection, and capital investment. It is likely that the colonies would be located near the necessary physical resources. It is possible that spaceflight may be transformed into a normality and that some Earthlings may migrate and establish colonies on Mars and possibly beyond.

In space settlements, a life support system must recycle or import all needed nutrients. The closest terrestrial analogue to space life support is that of a nuclear submarine. Nuclear submarines use mechanical life support systems to support humans for months without surfacing, and this same basic technology could possibly be employed for space use. However, nuclear submarines run open loop extracting oxygen from seawater, and typically dumping carbon dioxide overboard, although they recycle existing oxygen.

136

Colonization of the Moon

For long-term sustainability, a space colony should be close to self-sufficient. Mining and refining the Moon's materials on-site for use both on the Moon and elsewhere in the Solar System could provide an advantage over deliveries from Earth, as they can be launched into space at a lower energy cost than from Earth due to the Moon's lower gravity. Large amounts of cargo may need to be launched into space for interplanetary exploration in the future, and the lower cost of providing goods from the Moon would be beneficial.

The Moon may play a role in supplying space-based construction facilities with raw materials. Microgravity in space allows for the processing of materials in ways impossible or difficult on Earth, such as "foaming" metals, where a gas is injected into a molten metal, and then the metal is annealed slowly. On Earth, the gas bubbles rise and burst, but in a zero gravity environment, that does not happen. The annealing process requires large amounts of energy, as a material is kept hot for an extended period of time (allowing the molecular structure to realign), and this too may be more efficient in space, as the vacuum drastically reduces all heat transfer except through radiative heat loss.

Exporting material to Earth in trade from the Moon is problematic due to the cost of transportation, which would vary greatly if the Moon were industrially developed. One suggested trade commodity is helium-3, which is carried by the solar wind and accumulated on the Moon's surface over billions of years, but occurs only rarely on Earth. Helium-3 might be present in the lunar rocky material that covers bedrock in quantities of 0.01 ppm to 0.05 ppm (depending on soil).

Helium-3 harvested from the Moon may have a role as a fuel in thermonuclear fusion reactors. It should require about 100 metric tons (220,000 lb.) of helium-3 to produce the electricity that Earth uses in a year and there should be enough on the Moon to provide that much for 10,000 years.

To reduce the cost of transport, the Moon could store propellants produced from lunar water at one or several depots between the Earth and the Moon, to resupply rockets or satellites in Earth orbit, or Mars.

Terraforming Venus

The main problem with Venus today, from a terraforming standpoint, is the thick carbon dioxide atmosphere. The ground level pressure of Venus is 91 atm (1,330 psi)[ccxiv]. Due to the greenhouse effect, the temperature on the surface is several hundred degrees too hot for any significant organisms. Therefore, all approaches to the terraforming of Venus include somehow removing almost all the carbon dioxide in the atmosphere.

In 1961 Carl Sagan, American astronomer and astrophysicist, proposed the use of genetically engineered algae to fix carbon into organic compounds[ccxv]. Although this method is still proposed in discussions of Venus terraforming, later discoveries showed that biological means alone would not be successful. Difficulties include the fact that the production of organic molecules from carbon dioxide requires hydrogen, which is rare on Venus. Because Venus lacks a protective magnetosphere, the upper atmosphere is exposed to direct erosion by the solar wind and has lost most of its original hydrogen to space. As Carl Sagan noted, any carbon that was bound up in organic molecules would quickly be converted to carbon dioxide again by the hot surface environment. Venus would not begin to cool down until after most of the carbon dioxide had already been removed.

On Earth nearly all carbon is sequestered in the form of carbonate minerals or in different stages of the carbon cycle, while little is present in the atmosphere in the form of carbon dioxide. Air is roughly 78% nitrogen, 21% oxygen, and 0.04% carbon dioxide. On Venus, the situation is the opposite (Venus' atmosphere is about 96% carbon dioxide). Much of the carbon is present in the atmosphere, while little is sequestered in the lithosphere. Many approaches to terraforming focus on getting rid of carbon dioxide by chemical reactions trapping and stabilizing it in the form of carbonate minerals.

138

If these surface minerals were fully converted and saturated, then the atmospheric pressure would decline, and the planet would cool somewhat. To convert the rest of the carbon dioxide in the atmosphere, a larger portion of the crust would have to be artificially exposed to the atmosphere to allow more extensive carbonate conversion. Bombardment of Venus with refined magnesium and calcium from off-world could sequester carbon dioxide in the form of calcium and magnesium carbonates. About 8×10^{20} kg of calcium or 5×10^{20} kg of magnesium would be required to convert all the carbon dioxide in the atmosphere, which would entail a great deal of mining and mineral refining. Large amounts of carbon dioxide could be removed from the atmosphere by high-pressure injection into subsurface porous basalt formations, where carbon dioxide is rapidly transformed into solid inert minerals.

Other potential methods for terraforming Venus include direct removal of the atmosphere, cooling planet by solar shades, cooling by placing reflectors in the atmosphere, or introduction of water.

Astronomical Distances and Distant Travel

There are at least two trillion other galaxies in the observable universe. The distances between galaxies are on the order of a million times farther than those between the stars. Since there appears to be a speed of light limit on how fast any material objects can travel in space, intergalactic travel would require voyages lasting millions of years, as time measured on the Earth.

Astronomical units, abbreviated AU, are a useful unit of measure within our Solar System. One AU is the distance from the Sun to Earth's orbit, which is about 93 million miles (150 million kilometers).

For many interstellar distances astronomers use light years. A light year is the distance a photon of light travels in one year, which is about 6 trillion miles (9 trillion kilometers, or 63,000 AU). A light year is how far you would travel in a year if you could travel at the speed of light, which is 186,000 miles (300,000 kilometers) per second.

Distances to Bodies in the Universe

Distance to the Moon – 239,000 miles
Distance to Sun – 93,000,000 miles (1 AU)
Nearest Sun, Proxima Centauri (Alpha Centauri System) – 267,000 AU (4.22 light years)
Distance to Canis Major Dwarf Galaxy – 25,000 light years
Distance to Andromeda Galaxy – 2.5 million light years
Distant Galaxes - 13.2 billion light years
Edge of Observable universe – 46.5 billion light years

If velocities can be obtained that begin to approach the speed of light, several challenges would be encountered, including particle collisions and the effect of such gravitational forces on human anatomy, if not compensated. Electromagnetic energy including light can move at the speed of light, but that is not the case for matter. The faster an object moves, an exponentially larger amount of energy is needed to speed it up, which is why travelling at light speed requires an infinite and impossible amount of energy.

Moving at half the light of speed, the voyage to the nearest star would take 16 years round trip. Due to time dilation effects the crew would experience less time during the trip. Such effects only become substantial as one approaches the speed of light. The slowing of time on board a ship moving at various speeds is designated by gamma. Actual travel times are longer, considering that deceleration must also be part of each journey, to slow down to a zero speed at the endpoints.

140

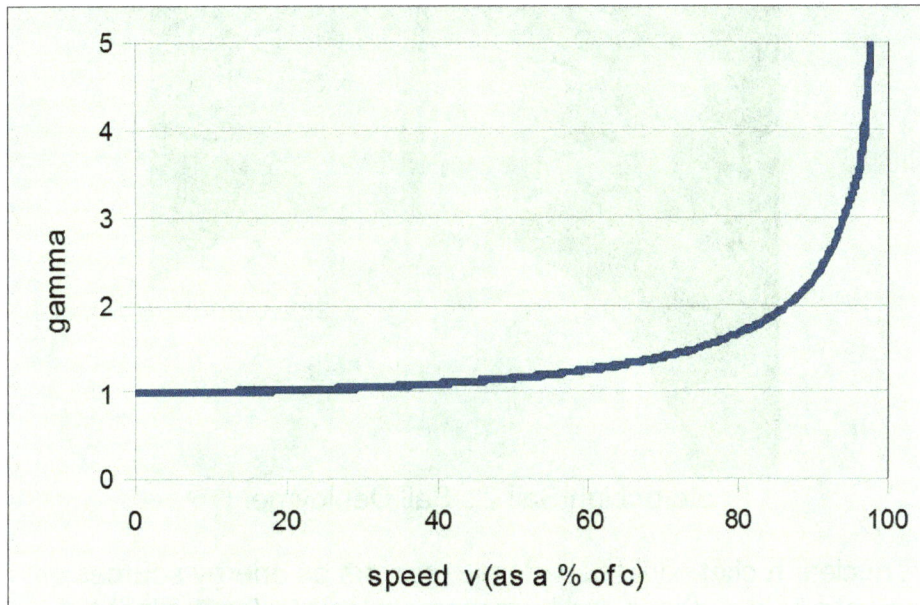

Time Dilation [Gamma] as Function of Speed [velocity] as % of c [speed of light]

Even at the speed of light it would take 25,000 years to reach our nearest galaxy, the Canis Major Dwarf Galaxy.

Many different technologies are being examined for potential interstellar transport including solar sails, high speed ion engines, nuclear rockets, and ramjet fusion. Solar sails could potentially use the pressure exerted by light using a gigantic solar sail. In 2019, the Planetary Society launched and deployed LightSail 2, which was released at an altitude of 720 kilometers. This effort is working toward the development of a solar sail that could potentially be used for propulsion.

Photo of LightSail 2's Sail Deployment[ccxvi]

A nuclear rocket could use atomic reactors as energy sources or potentially use atomic and hydrogen explosions (controlled) to propel the starship. A starship using ramjet fusion would scoop hydrogen gas and compress the gas into helium, releasing large quantities of energy. Expelling ionized matter instead of combustible gasses results in higher velocities of both propellants and spaceships, so ion engines will improve travel times over today's rockets.

Chapter 4 – The Distant Future

When we consider the distant future, we will take a geological perspective and an astronomical perspective of time.

Geological Perspective

From a geological perspective, we have seen that the Earth has changed over time due to climatic shifts and tectonic plate movements. These forces have impacted the evolution of life, which began around 3.5 billion years ago. Also, the temperature of the Earth has varied significantly over the past millions of years. We have seen a series of major extinctions over the past 550 million years, which are likely to be repeated as various environmental changes occur due to a variety of potential causes. These new extinctions are likely to make way for the rise of different species.

We also discussed that geothermal energy, which powers the tectonic plate movements, also generates the Earth's magnetic field, and heats the interior of the Earth. Since it relies on radioactive decay of elements, this energy is limited over geological time. We will look at when this resource will potentially be depleted and the harmful effects on the Earth and any life on Earth.

Climatic Changes

We have seen that the Earth's temperature has changed drastically over time from molten, to ice covered, to globally tropical, with cyclical periods of glaciation. Although we are technically in an ice age (ice on polar caps), we appear to be moving rapidly toward an increase in temperature of the Earth, largely due to the increase in greenhouse gases such as carbon dioxide. This will lead to melting of ice, rising sea levels, and changes in climates, which will cause some animals and plants to become extinct.

Since we know that the Earth's crustal movements have an impact on temperature as well as atmospheric content and continental positioning, which affects ocean currents, we can expect periods of climate change. There are likely to be a series of temperature rises and falls moving into the distant future, when the evolution of the Sun will have a large impact on climate change, through changes in the light luminosity and energy that strikes the Earth.

As discussed previously, the impact of a large asteroid or comet with the Earth could cause significant climatic change, major extinctions, or the destruction of life depending on the size and impact of the asteroid. The asteroid would form a crater and fill the atmosphere with dirt blocking sunlight, which would lead to colder temperatures while the particles are airborne. Later, the effects of crushed carbonate rock, blocked photosynthesis, and fires may increase in greenhouse gases, which would raise temperature. The asteroid could also cause damage to the ozone layer of the Earth, potentially leading to increased rates of cancer[ccxvii].

Also, the eruption of a mega volcano could darken the sky for years, causing temporary cooling of the Earth. The flood basalt resulting from the volcanic eruption or eruptions could cover large stretches of land or the ocean floor with basalt lava. Both effects will make life once again difficult for many species, especially large ones such as humans.

Tectonic Plate Movement

As previously discussed, due to the convection of the asthenosphere (upper layer of the earth's mantle) and lithosphere (rigid outer part of the earth, consisting of the crust and upper mantle), tectonic plates move relative to each other at different rates, from one to six inches (2.5-15 cm) per year. Over millions of years this movement becomes significant (1 million inches = 15.8 miles or 25.4 km). Researchers have developed a model of the Earth to predict the movement of one plate relative to another. The model, called MORVEL[ccxviii] for mid-ocean ridge velocities, describes the relative movements of the 25 interlocking tectonic plates that cover almost the entire surface of the Earth[ccxix].

144

It is difficult to predict accurately the future tectonic plate movements. It appears that the Atlantic Ocean will continue to widen, while the Pacific Ocean will decrease in size, but this could reverse at some time. It is anticipated that in ten to twenty million years the Atlantic will have widened by several hundred miles and the Pacific Ocean will have shrunk an equal amount. Australia will have moved north toward South Asia, and Antarctica will have moved slightly away from the South Pole in the direction of South Asia. The collision of India with Asia is causing the Himalayas to increase in magnitude, but this may not continue. The Mediterranean Sea may become smaller as Africa and Europe collide.

Movement of the plates can form subduction zones (one plate moving over another), generate mountains, and affect ocean currents. Changes in ocean currents combined with land movement relative to the equator can have effects on the climate.

Given the present trajectories across the globe, it appears that all the continents are headed for another collision. In approximately 250 million years from now most of the Earth's land will again form one giant supercontinent, which is called Pangaea Proxima.

Pangaea Proxima is a possible future supercontinent configuration that could occur within the next 300 million years (see below).

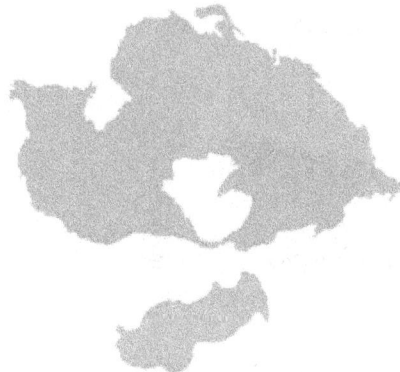

Pangaea Proxima[ccxx]

Evolution

Due to the high rate of extinctions that have been and continue to occur, it appears that we are currently in a sixth of a series of mass extinctions, since the beginning of life on the earth. This mass extinction will likely lead to the development of new species that will adapt and thrive under the changed environmental conditions. Humans have been involved in breeding for genetic characteristics and are currently involved in genetic engineering that will continue to affect the number of species and the genetic characteristics of those species.

Many species will adapt, evolve, and become new species or otherwise become extinct. Some species have been able to adapt or survive under the various environmental changes that have occurred over large periods of time. Examples include cyanobacteria (at least 2 billion years), ctenophora (700 million years), sponges (600 million years), jellyfish (at least 500 million years), velvet worms (500 million years), nautilili (500 million years), and horseshoe crabs (445 million years)[ccxxi]. We have previously discussed many species that have become extinct.

Ctenophore[ccxxii]

If humans do not become extinct, then a new species may evolve from the present human species.

Changes to humans may include an increase in height. Other changes could include smaller teeth, fewer toes, and weakened

146

muscle strength.[ccxxiii] Mono-ethnicity may be generated by the constant mixing of different ethnic groups over thousands of years.

Human height has increased over the last 200 years, possibly in part due to improved nutrition. In the last 150 years, the average human height has increased an average of around 3.9 inches[ccxxiv]. This trend appears to be leveling off and it could halt depending on dietary conditions in the future. Human teeth have decreased size by almost half over the last 100,000 years. There are millions of people today who are born without little toes and the need for a fifth toe appears minimal. Due to our increasing reliance on technology and machinery to do heavy work, there is a possibility of humans experiencing muscle atrophy. Since humans are globally mobile and mating with each other, over the course of millions of years, it is possible that humans will eventually evolve into a single ethnic group.

To determine what human evolution may lead towards, one would want to understand the forces that are leading to evolution. In this case, we would look for characteristics that lead to survival or advantage to survive. Currently, humans have few challenges to living to the age of reproduction. In the future, there may be shortages of food, water, energy, and possibly the need to survive in locations outside of the Earth. These trends might favor certain human traits. It is difficult to predict what human evolution may look like. Humans are likely, inadvertently or deliberately, to alter their genomes, altering their own evolution.

Some futurists predict a merging of humans with other biological or physical technology, which is already occurring due to advances in medicine, including implants and artificial organs, but it is debatable whether this should be called human evolution or human intervention.

It is reasonable that species such as rats, cockroaches, bats, etc., might do well in a future with limited resources but with available waste products from human consumption[ccxxv]. Advantageous traits that are beneficial to these and other animals as well as other types of life will likely be passed on to their progeny, leading to genetic changes in these life forms. Pets such as dogs and cats

147

almost certainly will continue to evolve with humans.

Due to evolution of the Sun, life on Earth during the next 500 to 1,000 million years would need to accommodate to a huge heat generating Sun[ccxxvi]. The equatorial regions of the Earth would likely be too hot for most life other than microbes. Existing plants would need to evolve to survive in a low-carbon dioxide environment. The oceans would be in the process of evaporating and huge salt flats would form along the shores. Most life in the sea would be gone except for those capable of adapting to the high salinity, such as some crustaceans. Much of the world would become desert with limited life forms such as lizards, snakes, scorpions, and cockroaches. It is possible that new species may evolve to adapt to the new environment. If humans were to continue to exist in this environment, they would likely have to live in the recesses of the Earth. There would be too much radiation from the Sun for humans to survive on the surface.

Advancements in technology may lead to what might be considered as evolutionary change or introduction of new entities. Some speculate that as the combination of robotics and artificial intelligence improves that we may be able to generate robots that are self-aware and are self-replicating. Efforts on general artificial intelligence could lead to conscious robots[ccxxvii]. Such entities could be helpful in many advancements, allowing humans to work on tasks best aligned with human skills. Such robots are also likely to be used in warfare. There is also some concern that these robots could eventually supplant humans.

If Earthlings move to other planets such as Mars or beyond, they may need to evolve over time to adapt to their new environments. Such adaptive forces could include differences in gravitational forces, breathing conditions (atmospheric conditions), temperature, radiation levels, etc. Also, higher levels of radiation on Mars or other locations would potentially lead to increased rates of genetic mutation that could accelerate evolution[ccxxviii].

Hundreds of thousands of years of biological and mechanical evolution may result in the beings different from current humans. These may be cyborgs (or cybernetic organism, beings both

148

biological and robotic).

It is possible that if we can visit other planets, we may find other life forms or possibly other species will encounter us. Given the billions of years of age of the Galaxy, it is puzzling that no aliens have yet been detected.

Types of Civilizations

In 1964, Nikolai Kardashev[ccxxix], Russian astrophysicist, proposed the idea that the status of a civilization depends on two primary things: energy and technology. He theorized that a civilization's technical advancement runs parallel to the amount of energy that the civilization can harness and manipulate. In recent years, scientists have expanded this scale to measure hypothetical civilizations—civilizations that are galactic, intergalactic, and even multiverse in nature.

A Type 0 civilization extracts its energy and raw-materials from crude organic-based sources such as wood, coal, and oil[ccxxx]. Any rockets utilized by such a civilization would necessarily require chemical propulsion. A civilization at this level would be largely confined to its home planet. This is about where we are.

A Type I civilization (10^{16} W energy usage), often known as a planetary civilization, can utilize and store all the energy available on its planet. Perhaps advances in technology such as energy generation using fusion could advance civilization to such a level. Michio Kaku, American theoretical physicist and futurist, tends to believe that we may reach Type I in 100 – 200 years. This means that we would need to boost our current energy production over 100,000 times to reach this status.

Type II civilizations (10^{26} W energy usage) dominate more than one solar system and/or are potentially capable of harnessing all the power available from a single star. A Type II civilization, often known as a stellar civilization, can use and regulate energy on a planetary scale. However, it cannot store this energy for future use. A Type II civilization can harness the power of their entire star (not merely transforming starlight into energy, but controlling the

149

star). Several methods for this have been proposed. The most popular method is that of the hypothetical Dyson Sphere. This device would encompass every single inch of the star, gathering most of its energy output and transferring it to a planet for later use. Alternatively, if fusion power (the mechanism that powers stars) has been mastered by the race, a reactor on an immense scale could be used to satisfy their needs. Nearby gas giants could be utilized for their hydrogen, slowly drained by an orbiting reactor.

A Type III civilization (10^{36} W energy usage), often called an interstellar civilization, can use, store, and regulate energy across multiple planets or star systems. However, it is not able to utilize or store any energy beyond its local universe. Even if we were to remain confined to Earth, a Type III civilization might still exist somewhere in the Universe. While theoretically possible, it is challenging to envision any civilization getting to this level.

Nikolai Kardashev believed a Type IV civilization was too advanced and did not go beyond Type III on his scale. Astronomers have extended the scale to Type IV (10^{46} W energy usage) and Type V (the energy available to this kind of civilization would equal that of all energy available in not just our Universe, but in all universes and in all timelines). These additions consider both energy access as well as the amount of knowledge to which the civilizations have access.

Geothermal Energy

As previously mentioned, heat is generated in the core of the Earth. There are several main sources of heat in the deep earth:

1. frictional heating, caused by denser core material sinking to the center of the planet
2. heat from the decay of radioactive elements

The deeper you go into the Earth, the higher the temperature. At 25 km down, temperatures rise as high as 750°C and at the core it is approximately 4000°C[ccxxxi]. Estimates of the total heat flow from Earth's interior to the surface span a range of 43 to 49 terawatts (TW).[ccxxxii] The two main sources in roughly equal amounts: are

150

the radiogenic heat produced by the radioactive decay of isotopes in the mantle and crust, and the primordial heat left over from the formation of Earth[ccxxxiii].

The Earth has been emitting heat for about 4.5 billion years and will continue to emit heat for billions of years into the future because of the ongoing radioactive decay in the Earth's core.

The radioactive decay of elements in the Earth's mantle and crust results in production of daughter isotopes and release of radiogenic heat. Four radioactive isotopes are responsible for the majority of radiogenic heat because of their enrichment relative to other radioactive isotopes: uranium-238 (^{238}U), uranium-235 (^{235}U), thorium-232 (^{232}Th), and potassium 40 (^{40}K). For the Earth's core, geochemical studies indicate that it is unlikely to be a significant source of radiogenic heat due to an expected low concentration of radioactive elements partitioning into iron.

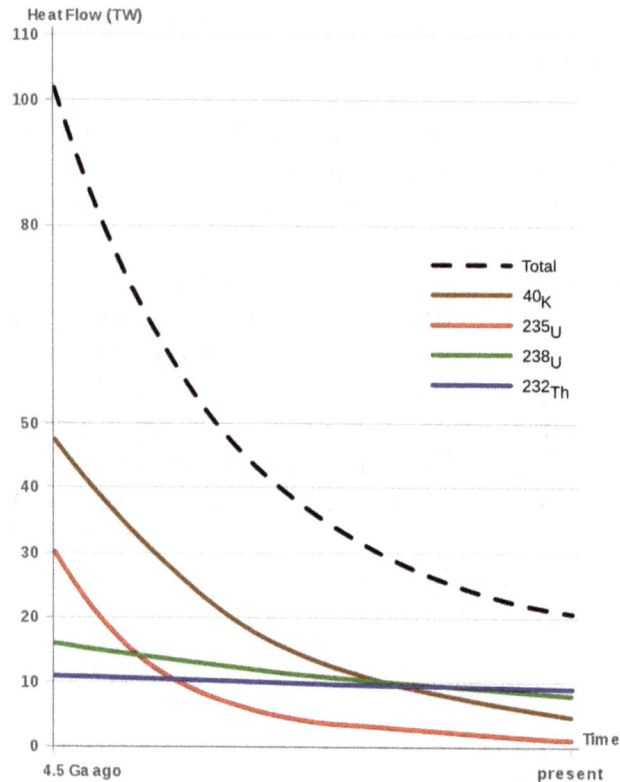

Radiogenic Heat Flow of Earth
(Giga-annum (Ga) = billion years ago)[ccxxxiv]

Primordial heat is the heat lost by the Earth as it continues to cool from its original formation. The Earth core's heat flow is thought to be due to primordial heat and is estimated at 5-15 TW (1 terawatt (TW) = 10^{12} W). Estimates of mantle heat loss range between 7 and 15 TW, which is calculated as the remainder of heat after removal of core heat flow and bulk-Earth radiogenic heat production from the observed surface heat flow.

We can expect a cooling of the Earth over time due to the loss of primordial heat (currently estimated at 4-15 TW) and the decrease in radiogenic heat to the decay of the radioactive isotopes (currently estimated at 20 TW). The half-lives of the ^{232}Th, ^{238}U, ^{40}K, and ^{235}U are in the range of 0.7-14 billion years. Thus, the total heat flow will slowly decrease in the next several billion years.

152

As the Earth cools this will affect the convective heating, which is responsible for tectonic plate movement and for generation of the geomagnetic field of the Earth. When the magnetic field collapses, the atmosphere can be stripped by the solar wind and flares.

Astronomical Perspective

From an astronomical time perspective, we will look over vast periods of time. We have seen the origin of the Universe around 13.8 billion years ago, which is expanding and continuing toward disorder. We also discussed the origin of stars, planets, and galaxies, and the changes to these bodies over time, which will continue. We have noted that eventual asteroid collisions are expected, and we will examine the potential merging of our galaxy (Milky Way) with the Andromeda galaxy. We also noted that our Sun has a limited energy source and that the Sun will evolve as this energy is depleted. When looking at astronomical time, we should consider that the Universe is young and that 13.8 billion years is small in comparison to the vastness of the future.

First, we will start with discussion of quasars, which have not been discussed previously.

Gamma Ray Burst

Gamma ray bursts (GRBs) are immensely energetic explosions that have been observed in distant galaxies. They are the most energetic and luminous electromagnetic events since the Big Bang. Gamma ray bursts are the most powerful explosions known in the Universe. They can lose as much energy as our Sun during its entire 10-billion-year lifetime in anywhere from milliseconds to a minute or more.

The blast generates two jets of gamma rays which travel out from the collapsing star in opposite directions at nearly the speed of light. Because their energy is strongly focused, the gamma rays emitted by most bursts are expected to miss the Earth and never be detected.

Gamma Ray Burst[ccxxxv]

The leading hypothesis is that GRBs originate from the merging of binary neutron stars or a neutron star with a black hole.

All GRBs observed to date have occurred outside the Milky Way galaxy and have been harmless to the Earth. However, if a GRB were to occur within the Milky Way within 5,000 to 8,000 light-years and its emission were beamed straight towards Earth, the effects could be harmful and potentially devastating for its ecosystems[ccxxxvi]. It could cause mass extinction or possibly even threaten life as we know it on our planet. Currently, orbiting satellites detect on average approximately one GRB per day.

Consequences of the Evolution of the Sun

Our Sun is in a process in which nuclear fusion reactions in its core fuse hydrogen atoms into helium. This process will eventually deplete some of the hydrogen. As the hydrogen is used up, the nuclear process will evolve, which will also have effects on the Sun.

154

The luminosity of the Sun is expected to steadily increase, resulting in a rise in the solar radiation reaching the Earth.[ccxxxvii] This will result in a higher rate of weathering of silicate minerals. Weathering of silicate rocks by carbonic acid is faster in a warmer climate because rainfall amounts tend to be greater. By providing calcium ions, weathering promotes limestone formation and removal of carbon dioxide from the atmosphere.

In about 600 million years, the level of carbon dioxide is expected to fall below the level needed to sustain carbon-3 fixation (taking carbon-3 from the atmosphere and putting it in organic molecules) photosynthesis used by trees. About 95% of shrubs, trees, and plants use carbon-3 fixation[ccxxxviii]. Some plants use the carbon-4 fixation method (examples include pineapple, corn, sugar cane), allowing them to persist at carbon dioxide concentrations as low as 10 parts per million. The long-term trend is for plant life to die off altogether. The extinction of plants will lead to the demise of almost all animal life, since plants are the base of the food chain on Earth.

In about one billion years, the solar luminosity is expected to be 10% higher than at the present. This will cause the atmosphere to become a moist greenhouse, resulting in a runaway evaporation of the oceans. No ice caps or glaciers will remain. Plate tectonics will likely come to an end, and with them the entire carbon cycle. Following this event, in about 2–3 billion years, the Earth's magnetic dynamo may cease, causing the magnetosphere to decay, leading to an accelerated loss of volatiles[ccxxxix] (chemical elements or compounds that can easily be vaporized such as nitrogen, carbon dioxide, ammonia, hydrogen, methane, sulfur dioxide) from the outer atmosphere. Four billion years from now, the increase in the Earth's surface temperature will cause a runaway greenhouse effect, heating the surface enough to melt it. By that point, all life on the Earth will be extinct.

The Sun will spend about another 5 billion years gradually becoming hotter because the helium atoms occupy less volume than the hydrogen atoms that were fused. As the core becomes denser, the Sun will become brighter. After about 5 billion years the Sun will start to turn into a red giant (expanded in size and the cooler outside turning redder). Our Sun is massive enough that the outward nuclear explosive forces will overcome the inward gravitation forces and the Sun will expand. Our Sun in about 7.5 billion years will have a radius close to the orbit of the Earth.[ccxl]

Image of what Earth may look like 5-7 billion years from now, when the Sun swells and becomes a Red Giant[ccxli]

Later the Sun will become a white dwarf (dense star the size of a planet with only the hot core remaining) and ultimately a black dwarf (white dwarf that has cooled such that it no longer emits significant heat or light). Regardless of the details and exact timing, the Sun will ultimately destroy all life as we know it on Earth.

Collision of Andromeda Galaxy and Milky Way Galaxy

If humans have escaped the Sun to colonize the Galaxy, another problem remains.

The Andromeda Galaxy is approaching the Milky Way at about 110 kilometers per second, as indicated by blueshift[ccxlii]. A Doppler blueshift means that the light received from Andromeda is getting closer to the earth.

There are several factors influencing the movement of galaxies. The Hubble flow, which is attributed to the expansion of the Universe, tends to make galaxies move away from each other.

156

The recession velocity is calculated as below.

$$\text{Recession Velocity} = H_0 \times D,$$

Where H_0 is the Hubble constant (approximately 70 km/sec/Mpc) and D is the distance in Mpc[ccxliii]. Megaparsec (Mpc) is one million parsecs or 3.26 million light years. There are also local flows and the motion of the galaxy with its cluster.

The Andromeda–Milky Way collision is a galactic collision predicted to occur in about 4.5 to 5 billion years between the Milky Way (which contains our Solar System including the Earth) and the Andromeda Galaxy. The stars involved are sufficiently far apart that it is improbable that any of them will individually collide. However, gas clouds will collide, potentially increasing radiation levels through bursts of massive star formation and supernova explosions. The Milky Way and Andromeda galaxies both contain supermassive black holes. The black holes may converge, which may cause some stars to be ejected from the resulting galaxy. The potentially resulting giant elliptical galaxy or super spiral galaxy is referred to as Milkomeda or Milkdromeda.[ccxliv]

Our local group, including the Milky Way, Andromeda, Triangulum and about 50 other, smaller galaxies are bound together. As time goes on, all the galaxies in our local group will merge, creating a giant elliptical galaxy, known as Milkdromeda. Beyond our local group, all the other galaxies, groups and clusters will continue to accelerate away from one another[ccxlv]. Thus, even if Earthlings survive these cosmic apocalypses, future Earthlings will be marooned the resulting giant local galaxy only to watch all other galaxies recede out of reach, eventually to disappear from view as the light from them fails to travel fast enough to be seen locally.

End of the Universe

The second law of thermodynamics states that total entropy of an isolated system can never decrease over time.[ccxlvi] The Universe is continually increasing in disorder. This means that while energy and matter are conserved, useful energy will be decreased to become useless equilibrated heat. A universe with equilibrated heat would approach a state of maximum entropy in which no further work is possible. Work is defined as a force causing the movement or displacement of an object. Without the ability to move or displace an object, life is not possible. Life is possible only because it is not isolated and has energy inflows. This state of maximum entropy is not expected to occur for about 1×10^{1000} years[ccxlvii]. This is an enormous amount of time, which is difficult to comprehend. Since this is such a long period of time from now, this concern is probably not as significant as a lot of other important events that will occur before this would impact Earthlings, if their descendants exist. By such a time, evolution would likely have changed our species to be largely unrecognizable.

From a long-term standpoint, as the entropy of the Universe continues to increase, life will ultimately be destroyed at any location. It appears that the Universe is expanding with an increased rate called the big rip, which would ultimately pull all matter apart. Even in the case of a cyclical Big Bang and collapse model, life will ultimately be destroyed. Albeit, in a cyclical model life could be regenerated after each Big Bang. There is currently a debate as to whether the Universe would tend toward a big rip (everything tears apart), big crunch (cosmos collapse), or big freeze (continued expansion)[ccxlviii]. The fate of our Universe depends on dark energy, which is generating an unexplained force that is counteracting gravity. As mentioned earlier, the Universe is expanding and if this continues then a big rip or big freeze would be expected, depending on the strength of the dark energy. If the strength of the dark energy diminishes, then a big crunch would be expected.

Conclusion

Below is a summary of past events, the current state with trending, and a summary of anticipated future events.

<u>Past Events – dates are approximate years ago</u>

13.8 billion - origin of the Universe
13.6 billion – origin of earliest stars and planets
 4.6 billion – origin of our Solar System
 4.5 billion – moon formed from impact with Theia
 3.5 billion – beginning of movement of the lithosphere due to
                tectonic plate movement
 3.4 billion – origin of life
650 million – Earth surface entirely or nearly entirely frozen
450 million – Ordovician-Silurian Extinction
375 million – Late Devonian Extinction
299 million – Permian-Triassic Extinction
201 million – Triassic-Jurassic Extinction
 66 million – Cretaceous-Paleogene Extinction
  6 million – appearance of humanoids
  2.6 million – use of stone tools
  2 million – early humans migrated out of Africa into Asia
200 thousand – human language
 75 thousand – Tova eruption
 40 thousand – use of fire
 12 thousand – beginning of farming and domestication of animals
  3.4 thousand – written language
  600 – scientific revolution
  300 – industrial revolution
  200 - medical revolution
  100 - electronic revolution
   65 - space revolution
   40 – digital revolution

<u>Current State and Trends</u>

- 5×10^{30} and 5×10^{31} bacteria and viruses on earth
- By mass – 82% and 12.7% of life are plants and bacteria

159

- Estimated to be 14 million species of which approximately 11 million are animals, 1.5 million are fungi, and 1 million are bacteria
- Human population is 7.79 billion
- Human population is increasing by 83 million annually, but trends show a slowing rate of increase
- Shortages of food, water, and energy are increasing
- Global warming is causing a variety of changes including higher temperatures and oceanic flooding and acidity
- Humans are transforming wilderness into agricultural land and are polluting water and air
- Many species are going into extinction
- Information is exploding in amount, content, and speed

Future Events

- Potential for global nuclear, biological, and chemical war
- Potential for pandemics
- Anthropocene Extinction and future extinctions
- Scientific advances including – improved agriculture, alternate sources of energy, improved manufacturing, advances in medicine, robotics, space exploration, computing
- Potential colonization of Mars
- Potential volcanic gas and dust disrupting agriculture
- Potential for asteroid impact
- Evolution of life forms including humans
- Continued tectonic plate movement
- Potential solar electromagnetic destruction of electronic devices

[Approximate years in future]
13,000 – Earth's axial tilt will be reversed[ccxlix]
20,000 – Chernobyl Exclusion Zone, 1000 square miles of Ukraine and Belarus becomes safe for human life
50,000 – Niagara Falls will have eroded away the 32 km to Lake Erie and will cease to exist

160

250,000 - Loihi, a young submarine volcano in the Hawaiian chain, rises above the Pacific Ocean's surface and becomes a new island

300,000? - Potential for gamma irradiation by a supernova event

1 million? – Supervolcanic eruption of magma comparable to the Toba supereruption 75,000 years ago

10 million - Widening East African Rift valley is flooded by the Red Sea, causing a new ocean basin to divide the continent of Africa and the African Plate into the newly formed Nubian Plate and the Somali Plate.

50 million – Current locations of Los Angeles and San Francisco will merge and the California coast will begin to be subducted into the Aleutian Trench

50 million – Africa's collision with Eurasia closes the Mediterranean Basin and creates a mountain range

250 million - All the continents on Earth may fuse into a supercontinent

600 million – loss of carbon dioxide in atmosphere, loss of some trees

1 billion – runaway evaporation of the oceans, no icecaps, loss of tectonic plate movement

2 billion – cooling of core of the Earth due to decrease in radiogenic heat, loss of geomagnetic field of Earth

4 billion – surface of Earth melts, no life on Earth

4.5 billion – collision of Andromeda and Milky Way Galaxies

5 billion – Sun is hotter and brighter and turns into a red giant

7.5 billion – Sun's radius is close to the orbit of the Earth

14 billion – Sun becomes a black dwarf

1×10^{1000} – thermal death of the Universe

We can observe small changes that are occurring even during our lifetimes. The lifetime of any creature is short in terms of comparison to the time involved in significant geological and astronomical changes. Much has happened during the first 13.8 billion years of our universe. A lot will happen in the billions of years of the future.

A favorite quote is of mine is by Deepak Chopra[ccl], Indian-born American author and alternative medicine advocate, "The past is history, the future is a mystery, and the moment is a gift. That is why this moment is called 'the present.'"

One observation of note is that there will never be another moment like this one in history. Change is continually happening. We should use our opportunities during the time that we have. We should learn from the past to help us in making good decisions in the present. We should make these decisions with a view toward the future, keeping in mind what is best for our future and for the benefit of those who come after us.

An interesting question is: Can humans escape the death predicted for the Earth?

We have seen that many species have become extinct and that many continue to go extinct. Often the reasons are associated with the inability to adapt to environment changes that were out of the control of these species. There are some circumstances that may be out of our control, such as being gamma irradiated by a quasar, but there are others that we may be able to moderate or avoid.

To escape the eventual effects of our Sun on the Earth, our descendants will need to be capable of space travel to move to another planet. While space travel within our Solar System appears probable, it is unclear whether this would be sufficient to allow sustainable habitation. It may be necessary to go outside of our Solar System to find a suitable environment. Technological advancements need to continue to develop to allow for space travel and colonization. The time required for distant travel, the negative effects of space travel on physiology, the challenges of energy and resources required will need to be overcome or mitigated.

Humans may become extinct by several means before the destruction of the Earth. Some potential causes include ecological, technological, and extraterrestrial.

162

Ecological

Climate change could result in human extinction. Continued destruction and pollution of our environment may lead to illness and infertility of humans. Humans could be eliminated by a pandemic that we might be unable to thwart. The extinction of animal and plant species may lead to disruption of the ecosystem, leading to increased pathogen transmission[ccli]. Increased biodiversity acts to decrease the spread of pathogens. For example, a lack of vultures can lead to a rise in other scavengers such crows, rats, and dogs, which may lead to increased incidence of disease including rabies[cclii].

Technological

If humanity allows self-destructive warfare to happen, including the use of biological warfare, it is possible that our capability for continued advancement may be stunted or destroyed. If this happens, humans or their descendants may not advance technologically to enable escape of Earth to another planet, where long term survival is possible.

Extraterrestrial

There is likely a lot that will happen in the next 2-4 billion years before humans, or the descendants of humans, need to leave the Earth because of our Sun's expansion. It is uncertain that humans or the evolutionary descendants of humans will continue to exist during that span of time as many catastrophic events are possible in the future.

Even if humans become extinct, it is likely that some life forms on Earth such as bacteria will continue to live for several billion years. If sufficient time is available, a subsequent life form (after human extinction) may be able to evolve from the existing life forms. Perhaps that new life form could develop the technology needed to colonize other planets. It is possible that an alien species could interact with Earthlings during the next few billion years.

After the destruction of all life on Earth due to the Sun's expansion or another cause prior to that, the Earth will continue to exist. The temperature, atmosphere, and environmental conditions will be very different as they currently are on Earth and these conditions will continue to change. As a function of the conservation of matter and energy, matter and energy will continue to exist in altered forms, as change continues to occur over the entire Universe.

Final Thoughts

The purpose of the book is to try to predict the future and see what the future may look like over time. As discussed at the beginning of the book, such predictions are necessarily limited by unpredictable events. Some may say that concerns associated today or for their lifetimes are sufficient to deal with. As I write this book, I am 62 years old and recognize that it is unlikely that I will live 40 years into the future, which is basically insignificant considering evolutionary time, geological time, or astronomical time. Regardless of this, I think it is important for us to consider the big picture of time as our collective actions over time will change the future. If we consider what the potential effects are of our actions and what we would like to see for the future, maybe we can act in a manner that will have a positive impact for the future of Earthlings.

Acknowledgements

Thanks to Marcia Bell, R.N. and Travis Bell, J.D. for reviewing early draft manuscripts. Margaret Gibbons, M.A.L.D, M.A. comments and suggested grammatical edits were very helpful. Thanks also go to Dr. Buss, Ph.D., high school classmate, for help with some of the technical content; Mike Nolin, professor in the Health Administration department at University of Maryland, Baltimore County; and Eric Bell, M.S.

Index

About the Author

Glenn Bell has a BS in electrical engineering from the University of Maryland, a MS in engineering from Johns Hopkins University, and a PhD in biomedical engineering from Louisiana Tech University. He has worked for Westinghouse Electric Corporation (Linthicum, MD), Kinetic Concepts (San Antonio, TX), Science Applications International Corporation, and Sequoia Software (Columbia, MD). He has taught chemistry (Louisiana Tech University), electrical engineering/technology (Capitol College) and physics (Carroll County Public School System). He is the author of scientific articles, papers, and patents. He enjoys exercise, reading, and writing. His previous book was "The Love of God is a Root of Evil" Lulu Press, Inc. 2020.

The opinions expressed in this book are specific to the author and are not the views of my past or present employers.

Endnotes

Introduction

[i] What are the odds of other intelligent life in the universe? - CBS News
[ii] 20120702-crystalball.jpg (240×157) (wordpress.com)

Chapter 1

[iii] https://en.wikipedia.org/wiki/Newton%27s_laws_of_motion
[iv] The origins of the universe facts and information (nationalgeographic.com)
[v] A Brief History of the Universe from Ancient Babylon to the Big Bang, J.P. McEvoy. Running Press 2010
[vi] 14_Origins-of-Solar-System.jpg (632×395) (wisc.edu)
[vii] https://en.wikipedia.org/wiki/Nuclear_fusion
[viii] hertzsprung-russell diagram - Bing images
[ix] http://www.columbia.edu/~vjd1/origins.htm
[x] Geosphere - Wikiversity
[xi] New Evidence for Protoplanet Theia Found in Moon Rocks | Space Exploration | Sci-News.com
[xii] https://www.sciencefocus.com/space/how-we-explained-the-origin-of-the-moon/
[xiii] How Did the Moon Form? - Universe Today
[xiv] Why the Moon is getting further away from Earth - BBC News
[xv] What Will Happen To Ocean Tides When The Moon Moves Away From Earth? » Science ABC
[xvi] File:Dynamo Theory - Outer core convection and magnetic field geenration.svg - Wikimedia Commons
[xvii] Plate tectonics - Wikipedia
[xviii] Supercontinent - Wikipedia
[xix] Pangaea - Wikipedia
[xx] https://www.nationalgeographic.com/science/earth/the-dynamic-earth/plate-tectonics/#close
[xxi] Evidence of Plate Motions - Geology (U.S. National Park Service) (nps.gov)
[xxii] main-qimg-34fd807aaeaaa8f1899344f549816dc5-c (480×360) (quoracdn.net)
[xxiii] LA's 'Big Squeeze' Continues, Straining Earthquakes (nasa.gov)

[xxiv] See a billion years of Earth plate tectonics movement in just 40 seconds - CNET

[xxv] 16.1 Glacial Periods in Earth's History – Physical Geology (opentextbc.ca)

[xxvi] Snowball Earth - Wikipedia

[xxvii] What Caused Snowball Earth? (Explained!) | Scope The Galaxy

[xxviii] 16.1 Glacial Periods in Earth's History – Physical Geology (opentextbc.ca)

[xxix] https://www.livescience.com/64813-milankovitch-cycles.html

[xxx] Widows to the Universe Image:/earth/climate/images/milankovich_lg.gif (windows2universe.org)

[xxxi] http://www.nbcnews.com/id/20393495/ns/technology_and_science-science/t/how-life-earth-began/#.WkkIYSOZPBI

[xxxii] Miller–Urey experiment - Wikipedia

[xxxiii] https://www.americanscientist.org/article/the-origin-of-life

[xxxiv] Thomas R. Cech: Exploring the New RNA World (nobelprize.org)

[xxxv] origin of RNA - Bing images

[xxxvi] History of Earth - Wikipedia

[xxxvii] charnia - Bing images

[xxxviii] The First Hominins – Hominoid1101

[xxxix] https://en.wikipedia.org/wiki/Impact_event

[xl] plate tectonics - Extinction | Britannica

xli https://earthhow.com/mass-extinctions/

[xlii] Figure_27_04_06.jpg (800×492)

[xliii] http://palaeos.com/paleozoic/ordovician/dapingian.html

[xliv] https://www.britannica.com/science/Ordovician-Period/Invertebrates

[xlv] https://www.nationalgeographic.com/animals/article/15911-metals-extinction-ocean-oxygen-ordovician-silurian

[xlvi] https://www.britannica.com/science/Devonian-Period

[xlvii] https://www.geovirtual2.cl/geoliteratur/palFraasGraptolites01.htm

[xlviii] echinoderm blastoid crinoid - Bing images

[xlix] https://www.britannica.com/animal/pelycosaur

[l] https://www.britannica.com/science/Permian-extinction/Alteration-of-the-carbon-cycle

[li] conanimals.jpg (350×350) (le.ac.uk)

[lii] aetosaurs - Bing images

[liii] Triassic–Jurassic extinction event - Wikipedia

[liv] CAMP Magmatism in the context of Pangea - Triassic–Jurassic extinction event - Wikipedia

[lvi] https://www.britannica.com/science/K-T-extinction

[lvii] https://astropeeps.com/2021/02/25/cretaceous-paleogene-extinction-the-chicxulub-event-and-the-fall-of-the-dinosaurs/

[lviii] https://astropeeps.com/2021/02/25/cretaceous-paleogene-extinction-the-chicxulub-event-and-the-fall-of-the-dinosaurs/

[lix] The Human Family's Earliest Ancestors | Science | Smithsonian Magazine

[lx] Introduction to Human Evolution | The Smithsonian Institution's Human Origins Program (si.edu)

[lxi] The first migrations out of Africa - The Australian Museum

[lxii] File: Putative migration waves out of Africa.png - Wikimedia Commons

[lxiii] Early human migrations - Wikipedia

[lxiv] The Story of How Humans Came to the Americas Is Constantly Evolving | Science | Smithsonian Magazine

[lxv] first arrival of homo sapiens into the americas - Bing images

[lxvi] When Did Ancient Humans Start to Speak? - The Atlantic

[lxvii] https://www.researchgate.net/publication/10734716_Descent_of_the_larynx_in_chimpanzee_infants

[lxviii] When did humans start wearing clothes? Discovery in a Moroccan cave sheds some light - CNN

[lxix] A Brief History of the Future, Jacques Attali, Arcade 2011

[lxx] The Development of Agriculture | National Geographic Society

[lxxi] History of writing - Wikipedia

[lxxii] Indus script - Wikipedia

[lxxiii] oracle bone script - Bing images

[lxxiv] When Did People Start Using Money? | Discover Magazine

[lxxv] Greatest Civilizations of All Time - Top Ten List - TheTopTens

[lxxvi] Who really invented the light bulb? - BBC Science Focus Magazine

[lxxvii] The Most Important Developments in Human History | Norwich University Online

[lxxviii] Energy sources through time – timeline — Science Learning Hub

[lxxix] 8 Oldest Known Planets in the Universe - Oldest.org

[lxxx] Fermi paradox - Wikipedia

[lxxxi] How Many Years Does It Take for a Star's Light to Reach Earth? (reference.com)

[lxxxii] 6 Mars: Evolution of an Earth-Like World | Vision and Voyages for Planetary Science in the Decade 2013-2022 | The National Academies Press (nap.edu)

[lxxxiii] Mars compared to Earth (phys.org)

[lxxxiv] Mars flyby - Wikipedia

[lxxxv] Water on Mars - Wikipedia

[lxxxvi] File:Valles Marineris PIA00178.jpg - Wikimedia Commons

[lxxxvii] File:Plates tect2 en.svg - Wikimedia Commons

[lxxxviii] List of largest volcanic eruptions - Wikipedia

[lxxxix] Climate Change and Global Warming: Global warming facts: What We Know (meterology.blogspot.com)

[xc] https://scied.ucar.edu/longcontent/predictions-future-global-climate

[xci] https://www.epa.gov/climate-indicators/climate-change-indicators-coastal-flooding

[xcii] https://www.climatehotmap.org/global-warming-effects/sea-level.html

[xciii]

https://images.search.yahoo.com/yhs/search;_ylt=AwrFF.e5awJj.AwkFDw2nIlQ?p=global+sea+level&ei=UTF-8&type=dhm_A0K4Y_ga_bsf__alt__ddc_srch_searchpulse_net&fr=yhs-domaindev-st_emea&hsimp=yhs-st_emea&hspart=domaindev&imgl=fsuc&fr2=p%3As%2Cv%3Ai#id=3&iurl=https%3A%2F%2Fupload.wikimedia.org%2Fwikipedia%2Fcommons%2Fthumb%2F6%2F65%2FTrends_in_global_average_absolute_sea_level%252C_1880-2013.png%2F693px-Trends_in_global_average_absolute_sea_level%252C_1880-2013.png&action=click

[xciv] The Sixth Extinction, Elizabeth Kolbert, 2014

[xcv] Environmental Monitor | Computer Model Predicts Climate Change into the Distant Future (fondriest.com)

Chapter 3

[xcvi] Which of the following is the best statement regarding the graph of atmospheric CO2 concentrations? - lifeder English

[xcvii] Carbon dioxide in Earth's atmosphere - Wikipedia

[xcviii] 5 Countries Producing Most Carbon Dioxide (CO2) (investopedia.com)

[xcix] https://commons.wikimedia.org/w/index.php?curid=688006

Chapter 2

[c] There Are More Viruses on Earth Than There Are Stars in the Universe | Daily Planet | Air & Space Magazine (airspacemag.com)

174

[ci] An Infinity of Viruses (nationalgeographic.com)

[cii] World Population Clock Live (theworldcounts.com)

[ciii] How many bacteria vs human cells are in the body? — The American Microbiome Institute

[civ] The Diversity of Life | Biology for Non-Majors II (lumenlearning.com)

[cv] gbo-ch-01-en.pdf (cbd.int)

[cvi] Let's get serious about protecting wildlife in a warming world (theconversation.com)

[cvii] World Population Clock: 7.8 Billion People (2021) - Worldometer (worldometers.info)

[cviii] Population growth - Wikipedia

[cix]
https://upload.wikimedia.org/wikipedia/commons/5/51/World_population_growth%2C_1700-2100%2C_2019_revision.png

[cx] https://ourworldindata.org/grapher/children-per-woman-un?tab=chart&time=1950..2015&country=OWID_WRL

[cxi] How Our Modern World Is Threatening Sperm Counts, Altering Male and Female Reproductive Development, and Imperiling the Future of the Human Race: Shanna H. Swan, PhD Scribner 2020

[cxii] https://ourworldindata.org/grapher/children-born-per-woman

[cxiii] The top 10 causes of death (who.int)

[cxiv] https://en.wikipedia.org/wiki/Mortality_rate

[cxv] https://www.pewresearch.org/wp-content/uploads/2019/06/FT_19.06.17_WorldPopulation_By-2100-five-of-10-largest-countries-projected-to-be-in-Africa.png?w=640

[cxvi] Energy and Food Security: Linkages through Price Volatility - ScienceDirect

[cxvii] Biodiversity — The Future Market

[cxviii] The worst famines in human history (scienceinfo.net)

[cxix] https://www.bbc.com/future/bespoke/follow-the-food/five-ways-we-can-feed-the-world-in-2050.html

[cxx] Water Scarcity | Threats | WWF (worldwildlife.org)

[cxxi] https://www.worldwildlife.org/threats/water-scarcity

[cxxii] https://www.theatlantic.com/international/archive/2012/05/the-coming-global-water-crisis/256896/

[cxxiii] https://www.livescience.com/49794-megadrought-prediction-southwest-plains.html

[cxxiv] DRIED UP: Lakes Mead and Powell are at the epicenter of the biggest Western drought in history | The Hill

cxxv 7.2 Water Supply Problems and Solutions – Environmental Biology (pressbooks.pub)

cxxvi global water shortage - Bing images

cxxvii Hybrid renewable energy systems for desalination | SpringerLink

cxxviii Land Use - Our World in Data

cxxix File:World electricity generation by source pie chart.svg - Wikimedia Commons

cxxx 1200-608637128-earth-atmosphere-layers.jpg (1200×900) (pixfeeds.com)

cxxxi Pollution Blamed for Thinner Air at Edge of the Atmosphere - The New York Times (nytimes.com)

cxxxii Space Junk from Satellites Polluting Earth's Exosphere | My World Through Rainbow-Colored Glasses (wordpress.com)

cxxxiii

cxxxiv ESS Topic 4.2: Access to Fresh Water - AMAZING WORLD OF SCIENCE WITH MR. GREEN (mrgscience.com)

cxxxv https://en.wikipedia.org/wiki/Blue_baby_syndrome

cxxxvi https://www.renewableresourcescoalition.org/pollution-causes-effects/

cxxxvii world land use map - Bing images

cxxxviii Land Use - Our World in Data

cxxxix The Lifespan of Common Plastic/Polymer Products (goecopure.com)

cxl number of people living in poverty - Bing images

cxli Global Education - Our World in Data

cxlii Literacy-rate - Literacy - Wikipediam.org

cxliii Global Education - Our World in Data

cxliv Global Education - Our World in Data

cxlv life span human over time - Bing images

cxlvi Occupational changes during the 20th century (bls.gov)

cxlvii https://news.yahoo.com/women-now-majority-u-workforce-215541650.html?fr=yhssrp_catchall

cxlviii Four Societal Changes Shaping The Future of Work - TalentCulture

cxlix https://teamstage.io/millennials-in-the-workplace-statistics/

cl http://www.pewresearch.org/fact-tank/2016/08/24/why-americas-nones-left-religion-behind/

cli Which Countries Use the Most Fossil Fuels? - Resource Watch Bloghttps://en.m.wikipedia.org/wiki/File:Religions_of_the_United_States.png, public domain

clii The Future of World Religions: Population Growth Projections, 2010-2050 | Pew Research Center (pewforum.org)
176

[cliii] Religiously unaffiliated people more likely to lean left, accept homosexuality | Pew Research Center

[cliv] https://www.thetimes.co.uk/article/post-christian-britain-arrives-as-majority-say-they-have-no-religion-5bzxzdcl6p3

[clv] Timeline of Religions • History Infographics

[clvi] VdagE.jpg (580×226) (imgur.com)

[clvii] United States Government: Why form a government? | United States Government (lumenlearning.com)

[clviii] Wealth Gap Graph - Bing images

[clix] Charts: Visualizing the Extreme Concentration of Global Wealth (visualcapitalist.com)

[clx] Climate change - Wikipedia

[clxi] https://www.e-education.psu.edu/geog885/sites/www.e-education.psu.edu.geog885/files/geog885q/file/Lesson_04/Predicting_Crime_groff.pdf

[clxii] https://www.nytimes.com/2022/08/10/world/europe/ukraine-drones.html#:~:text=For%20years%2C%20the%20United%20States%20has%20deployed%20drones,in%202020.%20But%20these%20were%20large%2C%20expensive%20weapons.

[clxiii] https://fr.wikipedia.org/wiki/Drone_de_combat

[clxiv] https://emerj.com/ai-sector-overviews/ai-drones-and-uavs-in-the-military-current-applications/

[clxv] incidence of war over history - Bing images

[clxvi] Most radiation-resistant lifeform | Guinness World Records

[clxvii] 10 Life Forms That Would Survive Nuclear War - Toptenz.net

[clxviii] Perspective: History of Pandemics (blacklistednews.com)

[clxix] https://en.wikipedia.org/wiki/Human_extinction

[clxx] https://astronomy.com/news/2021/03/asteroid-dust-found-at-chicxulub-crater-confirms-cause-of-dinosaurs-extinction

[clxxi] Asteroid shock: NASA warns of '100 percent' chance of asteroid impact | Science | News | Express.co.uk

[clxxii] arc-1991-ac91-0193-95ebc6-640.jpg (640×427) (picryl.com)

[clxxiii] https://www.thoughtco.com/why-it-matters-when-species-go-extinct-1182006

[clxxiv] https://pestsmart.org.au/toolkit-resource/economic-and-environmental-impacts-of-rabbits-in-australia/

[clxxv] Extinction Over Time | Smithsonian National Museum of Natural History (si.edu)

[clxxvi] Glyptodont - Wikipedia

[clxxvii] BBC NEWS | Science/Nature | Stopping the next extinction wave

[clxxviii] The Future of Life, Edward Wilson, Random House 2002

[clxxix] IUCN Red List of Threatened Species

[clxxx] Biodiversity, Species Loss, and Ecosystem Function | Sustainability: A Comprehensive Foundation (lumenlearning.com)

[clxxxi] https://www.iucn.org/resources/conservation-tool/iucn-red-list-threatened-species

[clxxxii] A Brief History of the Future, Jacques Attali, Arcade 2011

[clxxxiii] Kurzweil Claims That the Singularity Will Happen by 2045 (futurism.com)

[clxxxiv] https://2020.igem.org/wiki/images/a/a8/T--Grenoble_Alpes--3Dprinter.jpg

[clxxxv] Is fake meat better for you, or the environment? (nbcnews.com)

[clxxxvi] Regional entrepreneur reinventing agriculture with robotic innovation | Advance Queensland | Queensland Government

[clxxxvii] Genetically modified food - Wikipedia

[clxxxviii] An overview of the last 10 years of genetically engineered crop safety research (pps.net)

[clxxxix] Automated Hydroponics (the-vital-edge.com)

[cxc] Thorium Reactor - Thorium Power Plant (nuclear-power.com)

[cxci] IEEE Spectrum, October 2021, China's Thorium Gambit: A Prototype Power Reactor Could Go On Line by 2030, p9

[cxcii] Could Fusion Clean Up Nuclear Waste? - IEEE Spectrum

[cxciii] Safety Concerns - Nuclear Fusion (weebly.com)

[cxciv] ITER Image Galleries

[cxcv] IEEE Spectrum, January 2022, "A Pinch of Fusion," p54-55

[cxcvi] Physics of the Future How Science Will Shape Human Destiny and Our Daily Lives by the Year 2100, Michio Kaku, Doubleday 2011

[cxcvii] technology-exhibition-expo-robot-preview.jpg (728×1092) (pickpik.com)

[cxcviii] Nature Does It Better: Biomimicry in Architecture and Engineering (autodesk.com)

[cxcix] Everything you need to know about plant-based plastics | National Geographic

[cc] Everything you need to know about plant-based plastics | National Geographic

[cci] Top 10 Greatest Medical Discoveries of All Time | HealthGuidance.org

[ccii] The Future of Medicine | Harvard Medical School

[cciii] Top 6 Robotic Applications in Medicine - ASME

[cciv] How Medical Robots Will Help Treat Patients in Future Outbreaks - IEEE Spectrum

[ccv] 13940319151357735458534.jpg (800×557) (tasnimnews.com)

[ccvi] What Is Bioprinting? » Science ABC

[ccvii] 6127848729_b77b3d0a11_b.jpg (1000×665) (staticflickr.com)

[ccviii] Biology of Aging | National Institute on Aging (nih.gov)

[ccix] 594px-Space_elevator_structural_diagram.svg.png (594×1024) (wikimedia.org)

[ccx] Asteroid Mining (mit.edu)

[ccxi] https://en.wikipedia.org/wiki/SpaceX

[ccxii] https://en.wikipedia.org/wiki/Human_mission_to_Mars

[ccxiii] mars space colony - Bing images

[ccxiv] Terraforming of Venus - Wikipedia

[ccxv] The Planet Venus (science.org)

[ccxvi] Planetary Society Deploys LightSail 2's Solar Sail. What Does the Future Hold For Solar Sails? - Universe Today

Chapter 4

[ccxvii] Long Term Effects of an Asteroid Impact on Earth (seattlepi.com)

[ccxviii] MRVL_abstrct.pdf (wisc.edu)

[ccxix] https://www.zdnet.com/article/predicting-the-movement-of-the-earths-tectonic-plates/

[ccxx] Pangaea Proxima - Wikipedia

[ccxxi] 10 Oldest Species in the World | Oldest.org

[ccxxii] ctenophore - Bing images

[ccxxiii] Check Out The 7 Probabilities of Future Human Evolution (thefutureworld.org)

[ccxxiv] Human Height - Our World in Data

[ccxxv] Improbable Destinies Fate, Chance, and the Future of Evolution, Jonathan B. Losos, Riverhead Books, New York, 2017

[ccxxvi] Future Evolution, Peter Ward, Times Books, New York 2001

[ccxxvii] The Forth Age Smart Robots, Conscious Computers, and the Future of Humanity Byron Reese Atria Books, 2018

[ccxxviii] Colonizing Mars Could Speed up Human Evolution | Astronomy.com

[ccxxix] Kardashev scale - Wikipedia

[ccxxx] A Brief Explanation of the Kardashev Scale: How Far Can Humanity Really Advance? (futurism.com)

[ccxxxi] Where does the Earth's heat come from? (phys.org)

[ccxxxii] Earth's internal heat budget - Wikipedia

ccxxxiii

https://en.wikipedia.org/wiki/Earth's_internal_heat_budget#cite_note-15
ccxxxiv The K/U ratio of the silicate Earth: Insights into mantle composition, structure and thermal evolution - ScienceDirect
ccxxxv

https://www.tasnimnews.com/en/news/2018/10/05/1845060/astronomers-capture-sonic-boom-from-unseen-gamma-ray-burst
ccxxxvi https://en.wikipedia.org/wiki/Gamma-ray_burst
ccxxxvii https://en.wikipedia.org/wiki/Future_of_Earth
ccxxxviii Explore The Major Difference Between C3 And C4 Plants (byjus.com)
ccxxxix Volatiles - Wikipedia
ccxl https://www.universetoday.com/108734/what-is-the-future-of-our-sun/
ccxli

https://commons.wikimedia.org/wiki/File:Red_Giant_Earth_warm.jpg, public domain
ccxlii Andromeda–Milky Way collision - Wikipedia
ccxliii Hubble Flow | COSMOS (swin.edu.au)
ccxliv Andromeda–Milky Way collision - Wikipedia
ccxlv The Limits of How Far Humanity Can Go In The Universe (forbes.com)
ccxlvi https://en.wikipedia.org/wiki/Second_law_of_thermodynamics
ccxlvii

https://en.wikipedia.org/wiki/Graphical_timeline_from_Big_Bang_to_Heat_Death
ccxlvii The Beginning to the End of the Universe: The Big Crunch vs. The Big Freeze | Astronomy.com

Conclusion

ccxlix 33 Things That Will Happen to Earth in The Next Trillion Years According to Scientists | Bored Panda
ccl "The past is history, the future is a mystery, and this moment is a gift. That is why this moment is called 'the present.'" -Deepak Chopra - Undefeated Motivation
ccli Humans are more at risk from diseases as biodiversity disappears - Scientific American Blog Network
cclii How the Current Mass Extinction of Animals Threatens Humans (nationalgeographic.com)

180